PRACTICAL
PROBLEMS *in*
MATHEMATICS
for INDUSTRIAL TECHNOLOGY

Delmar's *PRACTICAL PROBLEMS in MATHEMATICS* Series

- *Practical Problems in Mathematics for Automotive Technicians, 4e*
  George Moore
  Order # 0-8273-4622-0

- *Practical Problems in Mathematics for Carpenters, 6e*
  Harry C. Huth
  Order # 0-8273-4579-8

- *Practical Problems in Mathematics for Drafting and CAD, 2e*
  John C. Larkin
  Order # 0-8273-1670-4

- *Practical Problems in Mathematics for Electricians, 5e*
  Herman and Garrard
  Order # 0-8273-6708-2

- *Practical Problems in Mathematics for Electronic Technicians, 5e*
  Herman and Sullivan
  Order # 0-8273-6761-9

- *Practical Problems in Mathematics for Graphic Artists*
  Vermeersch and Southwick
  Order # 0-8273-2100-7

- *Practical Problems in Mathematics for Health Occupations*
  Louise M. Simmers
  Order # 0-8273-6771-6

- *Practical Problems in Mathematics for Heating and Cooling Technicians, 2e*
  Russell B. DeVore
  Order # 0-8273-4062-1

- *Practical Problems in Mathematics for Industrial Technology*
  Donna Boatwright
  Order # 0-8273-6974-3

- *Practical Problems in Mathematics for Manufacturing, 4e*
  Dennis D. Davis
  Order # 0-8273-6710-4

- *Practical Problems in Mathematics for Masons, 2e*
  John E. Ball
  Order # 0-8273-1283-0

- *Practical Problems in Mathematics for Welders, 4e*
  Schell and Matlock
  Order # 0-8273-6706-6

Related Titles

- *Fundamental Mathematics for Health Careers, 3e*
  Hayden and Davis
  Order # 0-8273-6688-4

- *Mathematics for Plumbers and Pipefitters, 5e*
  Smith, D'Arcangelo, D'Arcangelo, Guest
  Order # 0-7061-x

- *Vocational-Technical Mathematics, 3e*
  Robert D. Smith
  Order # 0-8273-6806-9

PRACTICAL PROBLEMS in MATHEMATICS
for INDUSTRIAL TECHNOLOGY

Donna D. Boatwright

Italy, Texas
Independent School District

DELMAR
CENGAGE Learning

Australia • Brazil • Japan • Korea • Mexico • Singapore • Spain • United Kingdom • United States

DELMAR
CENGAGE Learning

Practical Problems in Mathematics for Industrial Technology
Donna D. Boatwright

Publisher: Robert D. Lynch

Editor: Mary Clyne

Production Manager: Larry Main

Art & Design Coordinator: Nicole Reamer

Cover Design: Dartmouth Publishing

For product information and technology assistance, contact us at
Cengage Learning Customer & Sales Support, 1-800-354-9706

For permission to use material from this text or product,
submit all requests online at **www.cengage.com/permissions**
Further permissions questions can be emailed to
permissionrequest@cengage.com

Library of Congress Control Number: 95-37205

ISBN-13: 978-0-8273-6974-0

ISBN-10: 0-8273-6974-3

Delmar
Executive Woods
5 Maxwell Drive
Clifton Park, NY 12065
USA

Cengage Learning is a leading provider of customized learning solutions with office locations around the globe, including Singapore, the United Kingdom, Australia, Mexico, Brazil, and Japan. Locate your local office at **www.cengage.com/global**

Cengage Learning products are represented in Canada by Nelson Education, Ltd.

To learn more about Delmar, visit **www.cengage.com/delmar**

Purchase any of our products at your local bookstore or at our preferred online store **www.cengagebrain.com**

Notice to the Reader
Publisher does not warrant or guarantee any of the products described herein or perform any independent analysis in connection with any of the product information contained herein. Publisher does not assume, and expressly disclaims, any obligation to obtain and include information other than that provided to it by the manufacturer. The reader is expressly warned to consider and adopt all safety precautions that might be indicated by the activities described herein and to avoid all potential hazards. By following the instructions contained herein, the reader willingly assumes all risks in connection with such instructions. The publisher makes no representations or warranties of any kind, including but not limited to, the warranties of fitness for particular purpose or merchantability, nor are any such representations implied with respect to the material set forth herein, and the publisher takes no responsibility with respect to such material. The publisher shall not be liable for any special, consequential, or exemplary damages resulting, in whole or part, from the readers' use of, or reliance upon, this material.

Printed in the United States of America
12 13 14 15 16   17 16 15 14 13

Contents

SECTION 1   INTRODUCTION / 1

SECTION 2 WHOLE NUMBERS / 5

Unit 1   Addition of Whole Numbers / 6
Unit 2   Subtraction of Whole Numbers / 12
Unit 3   Multiplication of Whole Numbers / 18
Unit 4   Division of Whole Numbers / 24
Unit 5   Review and Combined Operations on Whole Numbers / 31

SECTION 3 COMMON FRACTIONS / 38

Unit 6   Addition of Common Fractions / 45
Unit 7   Subtraction of Common Fractions / 51
Unit 8   Multiplication of Common Fractions / 57
Unit 9   Division of Common Fractions / 63
Unit 10   Review and Combined Operations on Common Fractions / 67

SECTION 4 DECIMAL FRACTIONS / 72

Unit 11   Significant Digits, Rounding, and Scientific Notation / 74
Unit 12   Addition of Decimal Fractions / 80
Unit 13   Subtraction of Decimal Fractions / 86
Unit 14   Multiplication of Decimal Fractions / 92
Unit 15   Division of Decimal Fractions / 97
Unit 16   Decimal and Common Fraction Equivalents / 102
Unit 17   Review and Combined Operations on Decimal Fractions / 110

SECTION 5 RATIO AND PROPORTION / 115

Unit 18   Ratios / 116
Unit 19   Proportion / 122
Unit 20   Combined Problems in Ratio and Proportion / 129

SECTION 6 PERCENTS, AVERAGES, AND ESTIMATES / 132

Unit 21   *Percent and Percentages / 133*
Unit 22   *Simple Interest / 141*
Unit 23   *Discount / 145*
Unit 24   *Averages and Estimates / 148*
Unit 25   *Review of Problems Involving Percents, Averages, and Estimates / 152*

SECTION 7 EXPONENTS AND ROOTS / 156

Unit 26   *Exponents and Order of Operations with Exponents / 157*
Unit 27   *Roots / 162*
Unit 28   *Review and Combined Operations on Exponents and Roots / 165*

SECTION 8 MEASUREMENT / 167

Unit 29   *Length and Angle Measurement / 170*
Unit 30   *Area and Pressure Measurement / 178*
Unit 31   *Volume and Mass Measurement / 183*
Unit 32   *Energy, Work, and Temperature Measurement / 188*
Unit 33   *Measurements Involving Time / 192*
Unit 34   *Review and Combined Problems on Measurement / 195*

SECTION 9 TABLES, CHARTS, AND GRAPHIC REPRESENTATION OF DATA / 199

Unit 35   *Line Graphs / 200*
Unit 36   *Pie Graphs / 208*
Unit 37   *Bar and Stacked Bar Graphs / 213*

SECTION 10 FORMULAS AND EQUATIONS / 218

Unit 38   *Representation in Formulas and Equations / 219*
Unit 39   *Solving Equations / 222*
Unit 40   *Formulas Common in Industrial Technology / 228*

SECTION 11 GEOMETRY AND TRIGONOMETRY / 231

Unit 41   *Pythagorean Theorem / 232*
Unit 42   *Trigonometric Functions / 237*

APPENDIX / *243*

Resistor Color Code / *243*
U.S. Customary System Measurement Equivalents / *243*
SI (Metric) System Measurement Equivalents / *244*
U.S. - Metric Equivalents / *245*
Formulas / *247*
Trigonometric Functions / *249*

GLOSSARY / *252*

ANSWERS TO ODD-NUMBERED PROBLEMS / *254*

Preface

Practical Problems in Mathematics for Industrial Technology is one in a series of widely used books. It is designed to give students, trainees, and others a review of basic mathematics principles and practical problem-solving experience covering a variety of topics in Industrial Technology. Each unit contains a short review of a math principle, worked out examples, and practical problems.

This book is designed to be a supplement to a mathematics or vocational mathematics text, which will provide more detailed discussion of mathematics topics. Several objectives guided the development of this text, including the development of problem-solving abilities, development of "number sense," and practice with real-world problems stated using terminology common to the areas of Industrial Technology.

Practical Problems in Mathematics for Industrial Technology has several features to help both students and instructors. Selected units contain problems marked with an exclamation point (!) by the problem number. These are "thought questions," which may require multiple steps or require a written response. Two achievement reviews are included in the instructor's guide as an effective assessment tool. A glossary is included to aid students with technical definitions. An appendix provides technical information needed in problems, measurement and conversion values, and formulas used in Industrial Technology applications. Answers to odd-numbered problems (except ! questions) are provided.

The text is designed for either calculator use or manual solutions. Instructions for calculator use (with algebraic logic calculators) are provided in appropriate units.

An instructor's guide contains answers to all problems, along with other instructional aids.

Donna D. Boatwright is the Technology Coordinator for the Italy Independent School District in Italy, Texas. She has taught Mathematics, Industrial Technology, and Computer Applications at the secondary and university levels, and has work experience in industry. She holds master's degrees in education and mechanical engineering.

Dedication

This text is dedicated to my husband Tommy, and our children Michelle, Michael, and Melanie Boatwright, who have provided tremendous support and encouragement during its development. A special "thank you" is extended to my parents, J T and Lou Dixon, and to Dr. Jerry Drennan, for their years of encouragement and example.

Introduction

The problems in this text can be worked either manually or by using a calculator. You will need to follow your teacher's instructions on whether calculators are to be used. Even if you use calculators, it is important that you understand the math operations and be *able* to work the problems manually.

ORDER OF OPERATIONS

Many mathematics problems involve more than one operation. Standard rules exist in mathematics which govern the order in which those operations are performed. Sometimes called the *Order of Operations* or *Algebraic Hierarchy*, they state that operations are performed according to the following "ranking":

1. Parentheses and other grouping symbols

2. Exponents and roots

3. Multiplication and/or Division (left to right)

4. Addition and/or Subtraction (left to right)

For multiplication and division, the operations should be performed from left to right as they occur in the problem, unless otherwise indicated by grouping symbols.

Many students find that a memory device is helpful in remembering the correct order. A common device for the Order of Operations emphasizes the first letters of the operations and uses the letters to start the words in the sentence "Please Excuse My Dear Aunt Sally." You or your teacher may know a similar memory device.

Basic Math Operations

The four basic math operations and the symbols commonly used for each operation include:

addition   +

subtraction   –

multiplication   × or *

division   ÷ or /

Note that the * and / symbols came into common use with the development of the computer since those characters are available on standard keyboards, while the × and ÷ symbols are more traditional.

Grouping Symbols

There are several types of grouping symbols used in mathematics. In the order in which they are normally used, they include parentheses ( ), brackets [ ], and braces { }. Grouping symbols can be "nested" by placing them inside other grouping symbols. The examples below show the use of grouping symbols.

Example: $2 \times (3 + 4) - 1$

$= 2 \times 7 - 1$ add inside the parentheses first

$= 14 - 1$ multiply before subtracting

$= 13$ subtract

Example: $8 \times \{[(33 + 1) \div 2] - (5 \times 3)\}$

$= 8 \times \{[34 \div 2] - 15\}$ work inside the parentheses

$= 8 \times \{17 - 15\}$ divide inside the brackets

$= 8 \times 2$ subtract inside the braces

$= 16$ multiply

(Note that both sets of parentheses can be simplified in the same step since they are not "nested.")

TYPES OF CALCULATORS

Calculators are common in education, business, and industry today. They are very useful in calculating the numerical values for solutions to mathematics problems. There are several types of calculators in use today, categorized by the way in which they process the math operations entered. Minor differences may occur between different brands of calculators in the exact sequence of steps to use for different operations, although they generally can be categorized as described below.

"Four-Function" Calculators

Many very inexpensive calculators (often called *four-function calculators*) process operations in the order in which they are entered. These calculators are not programmed to use the Order of Operations or Algebraic Hierarchy. When the sequence of operations **2 + 3 x 4 + 5** is entered, the answer displayed will be **25**, which is incorrect according to mathematics principles.

Algebraic Logic Or Direct Algebraic Logic Calculators

Most calculators today are programmed to perform operations according to algebraic logic. When an operation such as addition is entered, the calculator stores that operation to see if it is to be completed before the next operation. For example, if the first operation entered is addition, and the second operation entered is multiplication, the calculator will complete the multiplication operation and then add the previously stored addition. For the problem **2 + 3 x 4 + 5**, an algebraic calculator will display an answer of **19**. In determining that answer, the multiplication (3 x 4) is completed first, then that product is added to the 2 already entered, and then the 5 is added.

Although most calculators use algebraic logic, there are still differences in the way numbers and operations should be entered between different *brands* of calculators. *It is very important that you read and understand the instruction manual for your calculator*, especially when entering more complex operations such as multiple operation expressions, trig functions, exponents, and scientific or engineering notation. Most algebraic logic calculators also have keys for grouping symbols, but may be limited in the number of sets of parentheses which can be "open" at a given time.

The **Calculator Use** sections in this text will refer to calculators which use algebraic logic or direct algebraic logic.

Reverse Polish Notation Calculators

Some advanced engineering or scientific calculators use Reverse Polish Notation (RPN). Typically, the main difference between this format and algebraic logic is in the order in which numbers and operations are entered. If your calculator displays an error message or gives a different answer when an expression like **2 + 3 x 4 + 5** is entered, look at the calculator or instruction manual to see if it is identified as an RPN calculator. If your calculator uses this type of logic, it is very important that you read your manual and understand how to enter mathematical expressions correctly.

 ## LEARNING ABOUT YOUR CALCULATOR

Enter the expression **2 + 3 x 4 + 5 =** and record the answer from the display. Use the information above to determine whether your calculator uses algebraic logic, Reverse Polish Notation or performs calculations in the order in which they are entered. You should then read the instructions for your calculator and learn how to enter data and various operations correctly. Selected sections of this text include an introduction to the use of an algebraic logic calculator for that type of problem.

Whole Numbers

In our number system with 10 digits (known as *base ten* or the *decimal system*), each digit in a number has a value determined by its location in the number. In a whole number, the right-hand digit has a place value known as "ones." The digits to the left have place values of "tens," "hundreds," "thousands," etc. An example is the number 1,385, which represents one thousand (1,000), three hundreds (300), plus eight tens (80), plus five ones (5).

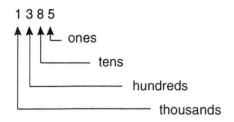

The concept of place value is important in operations on whole numbers. Numbers must be regrouped or renamed when performing certain mathematical operations. It is important to write neatly and write digits in the proper place value location to avoid mistakes when working with whole numbers.

Unit 1 Addition of Whole Numbers

BASIC PRINCIPLES OF ADDITION OF WHOLE NUMBERS

Addition is a mathematical process which produces a *sum*, or total value. When whole numbers are added, the numbers are normally written in a vertical "stack," with the place values lined up in columns. Write the numbers neatly to help avoid mistakes.

Each column of numbers is added, beginning with the right side. Write the last digit of the sum under the column. The remaining digit (if any) is carried to the next column and added. As an example, if the sum of the right column is 17, the 7 is written in the "ones" column and the 1 is moved to the "tens" column, since it actually represents 1 "ten".

Continue until all columns have been added. (Hint: If you are adding long columns, it may help to divide the numbers to be added into smaller groups, find a subtotal for each group and then add the subtotals to find the total.)

Example: Find the sum of 138 + 15 + 127 + 8 + 66.

3	*13*	*13*
138	138	138
15	15	15
127	127	127
8	8	8
66	66	66
4	54	354

KEY WORDS IN ADDITION PROBLEMS

There are several key words or phrases which occur frequently in problems for which addition is a correct solution strategy. Examples include *added to, plus, the sum of, more than, increased by,* and *total.*

 ## CALCULATOR USE

The addition key on most calculators is marked with a plus sign [+]. For calculators which use algebraic logic, enter the first number, press the [+] key, enter the next number, press the [+] key, until the last number is entered. The equal key [=] can be pressed to obtain the total. Be sure that you have read the instruction manual for your own calculator.

PRACTICAL PROBLEMS

1. 27 + 55 _____

2. 43 + 271 + 29 _____

3. 195 + 42 + 535 _____

4. 1,085 + 74 + 951 + 820 _____

5. 686 + 12 + 485 + 2,781 _____

6. 4,885 + 2,862 + 1,033 + 86 + 2,212 _____

7. 20,388 + 744 + 895 + 94 _____

8. 2,002 + 848 + 935 + 137,884 _____

9. 2,776 + 392 + 498 + 362 + 38 + 12 _____

10. 8,004 + 3,142 + 32,259 + 518 _____

Note: The symbol " refers to inches, while the symbol ' refers to feet. Both will be used throughout this text.

11. Assembling a printer housing requires twelve #10-24 UNC, four ¼-20 UNC, and two #6-32 UNC machine screws. What is the total number of machine screws in the assembly? (Discover: What is the meaning of the number before the dash in the description of a screw (the ¼ and #6)? What is the meaning of the number after the dash (20 and 32)?) _____

12. An inventory of the endmill stock in a machine shop showed 36 endmills of 1" diameter, 22 endmills of ⅝" diameter, 9 endmills of ¾" diameter, and 18 endmills of ⅞" diameter. What is the total number of endmills in the inventory? _____

13. An assembly used on a helicopter has four components. Part A weighs 105 pounds, Part B weighs 32 pounds, Part C weighs 27 pounds, and Part D weighs 9 pounds. What is the total weight of the assembly? _____

14. Part of a drafter's weekly timesheet is shown. Find the number of hours he spent this week working on the:

a. handle drawing   _____

b. assembly drawing for the single-handle faucet   _____

c. both drawings combined   _____

Name: _____   Dept. _____

Date	Drawing	Start	End	Hours
8/12/95	Chrome handle 2146	8:00	11:00	3
8/12/95	Chrome handle 2146	12:00	5:00	5
8/13/95	Chrome handle 2146	8:00	10:00	2
8/13/95	Assembly for s/h fct.	10:00	12:00	2
8/13/95	Assembly for s/h fct.	1:00	5:00	4
8/14/95	Chrome handle 2146	8:00	12:00	4
8/14/95	Assembly for s/h fct.	1:30	5:30	4

15. A plastics injection molding company recorded the number of occurrences of an incomplete fill in the molds, which resulted in rejected parts. The values for January through June are shown. How many total parts were rejected for that reason during the six-month period?   _____

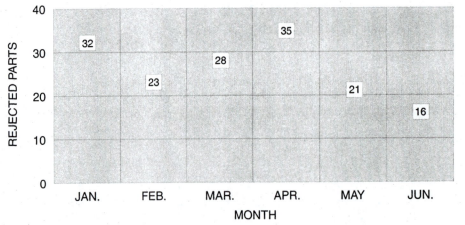

INCOMPLETE FILL REJECTS
INJECTION MOLDING DEPARTMENT

16. A warehouse received a shipment of 21,000 linear feet of ½" O.D. aluminum tubing from the manufacturer. The warehouse had orders from customers for the following amounts: 8,200 feet, 650 feet, 2,050 feet, 775 feet, 5,400 feet and 1,075 feet.

    a. What is the total length of tubing needed to fill the customers' orders? _____

    b. Did the warehouse receive enough in the shipment to fill all the orders? _____

17. A technology education supply company offers free shipping on any order totalling $1,000 or more. Part of the order form submitted by an Industrial Technology teacher is shown.

    a. What is the total cost of the items? _____

    b. Does his order qualify for free shipping? _____

To: XYZ Technology Supply
        P.O. Box 124
        Anytown, USA

P.O. Number 5130

Qty.	Size	Description	Unit Cost	Total Cost
3		Robot kit w/o computer interface	$ 52.00	$ 156.00
3	36" dia.	Dome kits	14.00	42.00
45		CO_2 Car kits	3.00	135.00
1	18×24	Economy tracing paper (500 sheets)	44.00	44.00
3	85 pc.	EZ-Set Modeling component kits	210.00	630.00
		Subtotal		
		Shipping/handling		
		Tax		
		Total		

18. An engineering technician compared purchasing a computer system as individual components rather than as a complete system. The prices of the components he selected were: CPU & keyboard $975, monitor $349, mouse $27, 3½" floppy disk drive $95, and printer $329. Another store has a sale on a complete system with the same components priced at $1695.

   a. What is the total cost of the computer purchased as individual units? _____

   b. Which is the better buy? _____

19. The line graph shows the monthly water consumption in gallons of a plating company. Find the total amount of water used in the 6-month period shown. _____

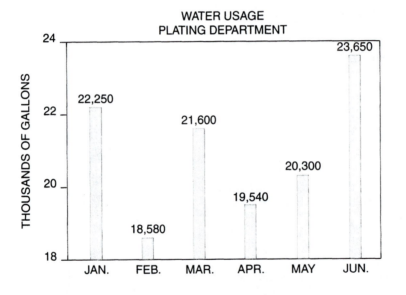

WATER USAGE
PLATING DEPARTMENT

20. A construction company used the following materials on a job: concrete and steel $1,150, framing lumber $472, hardware $65, roofing materials $380, wiring and electrical supplies $242, drywall supplies $112, paint and caulk $78, and flooring $432. What is the total cost of the materials used? _____

21. A courier service has a fuel storage tank which holds 2,500 gallons. During one week, the following amounts were pumped from the tank into trucks: 42 gallons, 63 gallons, 128 gallons, 103 gallons, 55 gallons, 61 gallons, and 130 gallons. How much fuel was pumped from the tank that week? _____

22. A kitchen has four appliances connected to a circuit: a coffeemaker (975 watts), a toaster (1,100 watts), a mixer (130 watts), and an electric skillet (1,225 watts). What is the total load when all appliances are operating? _____

23. The computer lab diagram shows the length of each connecting network cable. What is the total amount of cable needed? _____

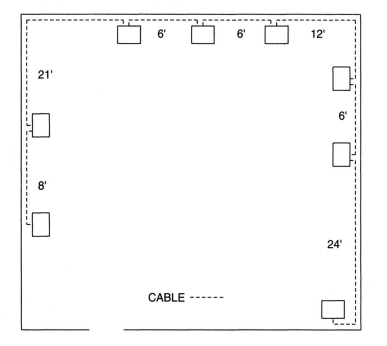

24. For resistors connected in series, the total resistance is the sum of the individual resistance values. A circuit contains four devices, connected in series, with resistances of 12 ohms, 140 ohms, 80 ohms, and 18 ohms. What is the total resistance in ohms? _____

25. An automotive service department is required to charge consumers a disposal fee when certain items are bought, with the fees sent to the state government. During March, the following fees were collected: tires $386, batteries $142, and motor oil $34. What is the total amount to be submitted? _____

Unit 2 Subtraction of Whole Numbers

BASIC PRINCIPLES OF SUBTRACTION OF WHOLE NUMBERS

Subtraction is the inverse (or "opposite") operation of addition. The answer to a subtraction problem is called the *difference* of the two numbers. The numbers should be aligned the same way that they are in addition, with the smaller value number placed below the larger value number.

Subtraction without Regrouping

Starting with the right column, subtract and write the digit below the column, as shown below.

Example: Subtract 978 – 524.

$$
\begin{array}{r}
9\,7\,8 \\
-\,5\,2\,4 \\
\hline
4\,5\,4
\end{array}
$$

Subtraction with Regrouping

If the digit being subtracted is larger than the top digit, regroup (or "borrow") 1 digit from the column to the left as shown in the example. This allows 1 ten to be rewritten as 10 ones, or 1 hundred to be rewritten as 10 tens, etc. When added to the value previously in the column, the resulting value is large enough for the subtraction to be done. The example below illustrates the regrouping process.

Example: Subtract 572 – 45.

$$
\begin{array}{r}
5\,7\,2 \\
-\,4\,5 \\
\hline
\end{array}
\qquad
\begin{array}{r}
6\ 12 \\
5\,\not7\,\not2 \\
-\,4\ 5 \\
\hline
5\,2\ 7
\end{array}
= 527
$$

The right column subtraction is 2 – 5, which cannot be completed using whole numbers. By regrouping one ten and adding ten ones to the right digit, the resulting subtraction is 12 – 5, which can be completed. *You may want to mark through and write the revised digits to help avoid mistakes, as shown in the example.*

KEY WORDS IN SUBTRACTION PROBLEMS

There are several key words or phrases which occur frequently in problems for which subtraction is a correct solution strategy. Examples include *minus, less than, decreased by, reduced by, subtracted from, deducted from, how many are left, remaining, exceeds by, profit, more than,* and *difference.*

CALCULATOR USE

The subtraction key on most calculators is marked with a minus sign [–]. For calculators which use algebraic logic, enter the first number, press the [–] key, enter the number to be subtracted, then press the [=] key.

If the math problem requires more than one operation or step, be sure that you enter the problem in the calculator correctly. One method involves entering the values in innermost grouping symbols first and "working your way out" of the grouping symbols. (You may want to use the [=] key to obtain subtotals in some cases.) Another method is to enter the expression as one sequence, using the parentheses keys as needed.

Example: If a storage tank contains 1,440 gallons of fuel, and 3 trucks are filled from the tank, using 68, 55, and 62 gallons respectively, how much fuel remains?

Solution 1: Add the amounts withdrawn from the tank and subtract that subtotal from the total fuel. Enter **68 + 55 + 62** and press = to get 185. Then, enter **1,440 – 185** and press = to get 1,255.

Solution 2: Enter the expression as one series of keystrokes, using grouping symbols. **1,440 – (68 + 55 + 62) =** to get 1,255. Entering the expression at once is normally the preferred method, especially when the calculations include other than whole numbers.

PRACTICAL PROBLEMS

1. 765 – 451 _____

2. 875 – 213 _____

3. 851 – 218 _____

4. 12,512 – 8,187 _____

5. 203,756 – 92,810 _____

6. 2,002 − (947 + 366) _____

7. 18,635 − [5,325 − (388 + 57)] _____

8. 2,002 − (947 − 366) _____

9. 9745 − [844 + (327 − 68)] _____

10. 27748 − [(3,764 − 1,212) + 84] _____

11. A CAD service was hired by a plumbing products company to scan 125 hard-copy drawings, converting them into a format which could be edited with a CAD program. If 87 of the drawings have been scanned, how many remain to be done? _____

12. In a machine shop, 3,450 pounds of 1½" diameter brass rod was purchased for a screw machine operation. When production was finished, 2,162 pounds of parts were packed for shipping, with the remaining brass sold to a recycler in the form of cleaned chips. What was the weight of the chips? _____

13. The warranty on Jim's new car requires that he have the oil changed after 3,000 miles. If his speedometer reading is 2,370, how many more miles can he drive before the oil change is due? _____

14. A telephone installer estimated that a job would require 155 feet of wire. He took a 200-foot roll to the job site. The actual amount used was 138 feet. How much wire was left on the roll? _____

15. A computer bulletin board allows 120 minutes of access per month for a $30 fee. A subscriber has accessed the service for the following times this month: 18 minutes, 36 minutes, 9 minutes, 24 minutes and 10 minutes. How many minutes are still available for use this month? _____

16. A chemical transport company has a fuel storage tank which holds 2,200 gallons. At the start of the week, the tank contained 1,850 gallons. During that week, the following amounts of fuel were pumped from the tank into trucks: 84 gallons, 63 gallons, 51 gallons, 55 gallons, 72 gallons, and 69 gallons. How much fuel remains in the tank? _____

17. The melting point of brass is 1,700°F, while zinc melts at 787°F. What is the difference in the melting temperatures? _____

18. An inventory of computer cable on July 1 showed 1,285 feet. During July, the lengths removed from inventory were 122 feet, 240 feet, 88 feet, and 355 feet. How many feet of cable are left in inventory at the end of the month?

     _____

19. On August 1, a computer store had in stock 240 boxes of 3½" DSHD floppy disks and 134 boxes of 5¼" DSHD floppy disks. The daily sales of both types of floppy disks for the next week are as shown. At the end of the month, how many boxes of each type of disk remain in stock?

     3½" disks _____

     5¼" disks _____

FLOPPY DISK SALES (DSHD)

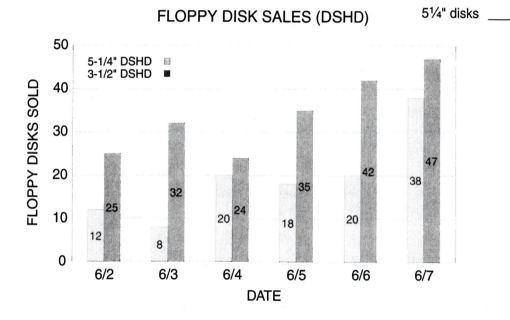

20. The Industrial Technology department at a vocational-technical college operates a recycling program with 5 collection sites. During August, they collected the following amounts of glass: 87 pounds, 112 pounds, 42 pounds, 84 pounds, and 92 pounds.

    a. Find the total weight of the glass collected.

     _____

    b. If their goal for August was 360 pounds, by how much did they exceed their goal?

     _____

21. A graphic design company charged $480 to photograph products for a company's advertising brochure. If the costs incurred were $24 for film, $216 for labor, and $42 for developing, what was the profit on the job? _____

22. A lumber yard had in stock 850 bundles of white 3-tab composition shingles. In the next five days, 25 bundles, 90 bundles, 51 bundles, and 76 bundles were removed from stock and sold. How many bundles remain in inventory? _____

23. A faucet manufacturing company reorders boxes for packaging its single-handle model when the supply reaches 400. If the inventory on March 1 was 1,426, with 360 boxes used on March 12 and 475 on March 20, how many more can be used before reordering? _____

24. A computer network installer took a 300-foot roll of cable and 45 connectors to a job. If he used 182 feet of cable and 28 connectors,

    a. how much wire is left? _____

    b. how many connectors are left? _____

25. A builder charged a customer $875 to build cabinets. The material cost was $310, hardware cost was $27, labor costs were $240, and finishing supplies cost $12. What is the profit on the job? _____

26. A school is planning to host a CO_2 car competition. Four schools will compete, with the number of entrants from each school registered as 15, 21, 7, and 14.

    a. If the host school has only 34 CO_2 cartridges in stock, how many more do they need to order in order to have one per student? _____

    b. If they want to have 20 extra cartridges, how many should they order? _____

!27. Kilowatt-hour meters, which measure energy used, have dials as shown below. To read the meter, start with the left dial and read each dial as you move to the right. (Note that the digits alternate reading clockwise and counterclockwise.) Record the last digit which the dial hand has reached or passed for each dial. The June reading was 35,428. Read the dials as they appeared on July 1 and determine how many kilowatt–hours of energy were used during the month.

_____

KWH METER DIALS

!28. On the meter below, the 2nd dial appears to be exactly on 5. Compare the two examples, and write a clear explanation of how you would know whether to record the 2nd dial as a 4 or 5.

_____

KWH METER DIALS

Unit 3 Multiplication of Whole Numbers

BASIC PRINCIPLES OF MULTIPLICATION OF WHOLE NUMBERS

Multiplication is a way of performing repeated addition more quickly. Numbers which are multiplied together are called *factors*. As an example, if three 5s are added, the sum is 15. If the number 5 is multiplied by 3, the answer or *product* is 15.

```
    5         5
    5       × 3
    5        15
   15
```

It is important that you know (or have memorized) the multiplication facts, at least through the 9's.

Multiplication of multi-digit numbers involves repeated use of single-digit multiplication. To multiply numbers with several digits, write the number to be multiplied. Below it, write the number of times it is to be multiplied, keeping the right column aligned.

Example: Multiply 235 × 57.

```
                                          2          12         1 2
      3        2 3       2 3       2 3       2 3        2 3
    2 3 5    2 3 5     2 3 5     2 3 5     2 3 5      2 3 5
    × 5 7    × 5 7     × 5 7     × 5 7     × 5 7      × 5 7
        5       4 5    1 6 4 5   1 6 4 5   1 6 4 5    1 6 4 5
                                      5         7 5    1 1 7 5
                                                       1 3 3 9 5
```

Multiply the numbers in the right column (7 × 5 = 35). Write the 5 in the ones column of the product (below the 7) and carry the 3 to the tens column. Then multiply (7 × 3 = 21) and add the 3 (21 + 3 = 24). Write the 4 beside the 5 in the product and carry the 2 to the next column. Multiply (7 × 2 = 14) and add the 2 that was carried (14 + 2 = 16). Since there are no more columns in the top number, write the 16 beside the previous digits in the answer. Next, multiply 235 by the 5 in 57. The answer digits should begin under the 5, rather than in the far right column. Multiply (5 × 5 = 25). Write the 5 in the next row of the answer, and carry the 2 to the next column. Multiply (5 × 3 = 15) and add the carried 2 (15 + 2 = 17). Write the 7 next to the 5 and carry the 1. Multiply (5 × 2 = 10) and add the carried 1 (10 + 1 = 11). Since there are no more digits in the top number, write the 11 beside the previous digits in the answer. The final step is to add the rows in the answer to obtain the product.

Multiplication Notation

When writing multiplication problems, there are three ways to indicate multiplication: using an "×" between the numbers, using a "dot" between the numbers, or placing the numbers in parentheses. Each of the following expressions represents the same mathematical value: 3×5; $3 \cdot 5$; $(3)(5)$.

Use the notation you prefer, but be sure that your mathematical representation of the problem clearly shows the operations to be performed.

KEY WORDS IN MULTIPLICATION

There are several key words or phrases which occur frequently in problems which use multiplication. Examples include *multiply, times, groups of, each, product of, multiplied by, of,* and *total.* Also, problems which contain phrases like *gallons per day* and *days* often require that those two values be multiplied to obtain the specified unit, like gallons.

CALCULATOR USE

The multiplication key on most calculators is marked with [×]. For calculators which use algebraic logic, enter the first factor, press the [×] key, enter the next factor and then the [=] key to display the product. If a series of factors are multiplied, you may enter all the factors, with the multiplication key pressed between factors, until all factors have been entered, and then press the [=] key.

If the calculations contain more than one math operation, be sure that you enter the values in a way that follows the Order of Operations or use parentheses as needed.

PRACTICAL PROBLEMS

1. 37×12 _____

2. 243×9 _____

3. $1,057 \times 42$ _____

4. 487×241 _____

5. $12,354 \times 54$ _____

6. $2,456 \times 44$ _____

7. $4,250 - (67 \times 3)$ _____

8. 68,208 – [37 × (367 – 289)]   _____

9. 12,458 × [(377 × 2) – (27 + 54)]   _____

10. 2,750 – [(330 × 4) + (27 × 5)]   _____

11. An engine is tested by running for 45 minutes at 3,650 rpm. How many revolutions does the crankshaft turn during that time?   _____

12. A CAD service charges $35 to scan a C–size print. If an architect has 24 C–sized prints scanned, what is the total cost?   _____

13. A transformer coil is wound with 25 layers, each having 88 turns. How many turns are on the coil?   _____

14. If a manufacturing company has four milling machines, each capable of producing 18 of a certain product per day, what is the maximum total daily output of all four machines?   _____

15. An engineering company leased 8 computer workstations for a 2-year period. If the monthly charge for each workstation is $125, what is the total cost of the lease?   _____

16. A company has 3 workers earning $7 per hour, 4 workers earning $8 per hour, and 2 workers earning $10 per hour. What is the total payroll if each worker works 40 hours?   _____

17. Ohm's Law states that there is a mathematical relationship between voltage (E), amperage (I), and resistance (R) in a circuit. It can be written as $E = I \times R$. For an electrical circuit with 15 ohms of resistance carrying 8 amps of current, what is the voltage for the circuit?   _____

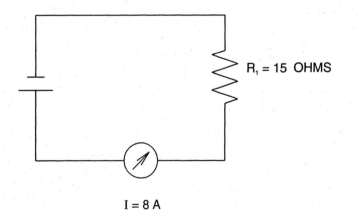

$R_1 = 15$ OHMS

$I = 8$ A

18. A manufacturing assembly requires twelve #10-24 UNC and four ¼-20 UNC machine screws. How many of each will be required for a production run of 2,250 assemblies?

    #10-24 UNC _____

    ¼-20 UNC _____

19. A stamping company uses 4,725 feet of 6" wide 18 ga. steel per day. Anticipating possible shipping problems from the supplier, the company wants to have in stock a 15–day supply. How many feet of steel will they need? _____

20. Sandpaper grit is sorted by passing it through screens with various sizes of openings. For example, 60-grit paper is sorted on a screen with 60 openings per linear inch, as shown. How many openings are there in a 1" by 1" section of screen for 60-grit paper? _____

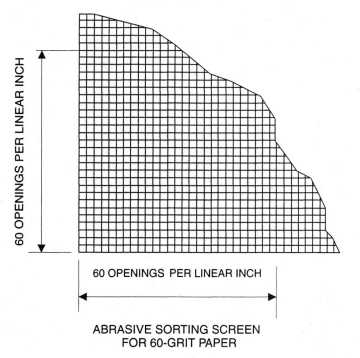

60 OPENINGS PER LINEAR INCH

ABRASIVE SORTING SCREEN
FOR 60-GRIT PAPER

21. A drafter worked for 6 days, 8 hours per day, on an exploded assembly drawing. If he earns $9 per hour, how much did the company spend on labor having the drawing prepared? _____

22. The contract for construction of a new building requires that the contractor pay a penalty for each day that the building is not finished after the scheduled finish date. If the penalty is $375 per day and construction requires an additional 12 days, what is the total amount of penalty to be paid?   _____

23. Foot-pounds are a unit of measurement for work in the U.S. Customary system. Work is defined as force × the distance over which the force is applied. If a motor weighing 45 pounds is lifted 4 feet during installation, what is the amount of work done (in foot-pounds) in lifting the motor?   _____

24. A maintenance agreement was purchased on the computer equipment in an engineering facility. The annual costs were $55 per computer, $62 per laser printer, and $58 per drum plotter. The lab contained 12 computers, 3 laser printers, and 2 drum plotters. What was the total annual cost of the contract?   _____

25. A stamping operation uses four presses, each capable of 1,080 stamping operations per hour. Each stamping operation produces three parts. If the presses operate for 8 hours each, what is the total number of parts produced?   _____

26. A buyer purchased 48 cartons of O-rings, each containing 12 packages of 80 rings each. How many O-rings were purchased?   _____

27. Review problem 20. How many openings are there in 1 square inch of the sorting screen for 220-grit sandpaper?   _____

28. Using Ohm's Law as described in problem 17, find the voltage for a circuit with 4 amperes of current and 55 ohms of resistance.   _____

29. The overhead cost for the finish sanding department in a furniture manufacturing company is stated as $28 per hour. If the department operates for 12 hours on Monday, 11 hours on Tuesday, 8 hours on Wednesday, and 10 hours each on Thursday and Friday, what is the total department overhead cost for that week?   _____

30. David, Mary, and Steven each earn $8 per hour as inspection trainees in an aircraft manufacturing plant. The number of hours each worked is shown on the weekly log sheet. Find the total earnings for each person and the total payroll for the trainees.

David _____

Mary _____

WEEK OF JULY 7–11

Steven _____

EMPLOYEE	M	T	W	T	F
DAVID	6	7	6	4	8
MARY	8	5	5	6	8
STEVEN	4	4	4	6	7

Total _____

31. An electric bill is based on kilowatt–hours of energy used. The reading for February 1 was 31,147 and on March 1, it was 34,209.

    a. How many kilowatt–hours were used during the month? _____

    b. If the charge for the first 1,000 kWh is 9.8¢/kWh and is 9.2¢ per kWh for the next 4,000 kWh, find the charge for energy used. _____

Unit 4 Division of Whole Numbers

BASIC PRINCIPLES OF DIVISION OF WHOLE NUMBERS

Division is a way of performing repeated subtraction quickly. For example, to determine how many times 23 can be subtracted from 74, you could actually subtract, as shown:

$$\begin{array}{r} 74 \\ -\ 23 \\ \hline 51 \\ -\ 23 \\ \hline 28 \\ -\ 23 \\ \hline 5 \end{array}$$

However, the process of repeated subtraction can be long and tedious. The same results could be obtained by division. The example above can be used to identify terms used in division. The number to be divided (74) is called the *dividend*. The number indicating how many times it is to be divided (23) is called the *divisor*. The divisor appears after the division sign when the problem is written as 74 ÷ 23 or as 74/23. The answer is known as the *quotient* (3) and sometimes can include a *remainder* (5). The standard way to write a division problem is shown below. The dividend is placed inside or under the division bracket, and the divisor is placed to the left, outside the bracket. The quotient is placed above the bracket. A remainder (if any) may be written to the right of the quotient.

$$\overset{\text{quotient}}{\text{divisor}\ )\ \overline{\text{dividend}}}\quad \text{remainder}$$

Example: Divide 74 by 23.

$$\begin{array}{r} 3 \quad \text{R 5} \\ 23\ )\overline{74} \\ \underline{69} \\ 5 \end{array}$$

Place the dividend (74) under the bracket. Place the divisor (23) to the left of the bracket. The number 23 will not divide into the first digit of the dividend (7), so it is divided into the first two digits (74). Using your knowledge of multiplication facts, select a number which, when multiplied by 23, will be close to 74, but not over 74. In this example, the number 3 is selected. The 3 is written above the last digit in the number divided (74). Multiply the 3 by 23 and write the product below 74.

Subtract 69 from 74 and write the difference below. In this case, there are no more digits in the dividend, so division can stop and the remainder can be rewritten next to the quotient.

If there were other digits in the dividend, the next digit would be brought down to the right of the 5, and the division steps would be repeated, as shown in the next example.

Example: Divide 740 by 23.

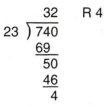

When performing division, keep in mind that it is the inverse operation of multiplication. You will need to know the multiplication facts and understand the forms in which they can be applied. For example, the fact 3 x 2 = 6 can be restated as 6 ÷ 2 = 3 (read as "six divided by two equals three") or as 6 ÷ 3 = 2. The forward slash (/) can also be used to indicate division, as in 6/3=2. *You should check your work on division problems by multiplying the quotient by the divisor and then adding the remainder; the result should be the dividend.* In the example problem, check by multiplying 32 by 23 and adding 4.

PRIME FACTORS OF A NUMBER

Remember that the term *factor* refers to numbers which are multiplied together. A prime factor is a number which cannot be divided evenly by any whole number other than itself and 1. The first five prime numbers are 2, 3, 5, 7, and 11. Numbers which are not prime are called *composite*. Examples include 4, 9, and 15. In some math operations, especially with fractions, it may be helpful to write a number as a product of prime factors. This is done by repeated division of the number by prime numbers until the last quotient is prime.

Example: Find the prime factorization of 54:

54 ÷ <u>2</u> = 27

27 ÷ <u>3</u> = 9

9 ÷ <u>3</u> = <u>3</u>

The prime factorization of 54 is thus 2 × 3 × 3 × 3 (the product of all the prime numbers used as divisors and the final quotient, underlined in the example). Note that there can be no remainder in this process.

KEY WORDS IN DIVISION PROBLEMS

There are several key words or phrases which occur frequently in problems involving division. Examples include *per, average, divided equally, each, for every,* and *groups of __ each*. The term "per," as in miles per gallon, indicates division of the first quantity by the second quantity (for example, miles divided by gallons).

Example: Each shipping carton used in a factory holds 15 faucet packages. If the total order for a warehouse is 630 faucets, how many cartons are needed?

$630 \div 15 = 42$

Example: The daily production runs for a cabinet assembly line were 42 on Monday, 37 on Tuesday, 31 each on Wednesday and Friday, and 54 on Thursday. What is the average daily production?

$(42 + 37 + 31 + 54 + 31) \div 5 = 39$

When finding an average value, add up the individual values and divide the sum by the *number* of individual values which were added.

CALCULATOR USE

The division key on most calculators is marked with a [÷] sign. For calculators which use algebraic logic, enter the dividend, then press the [÷] key, then enter the divisor, and press the [=] key. The quotient will be displayed. If a decimal point appears, followed by other digits (as in 32.45), the division involved a remainder or did not divide "evenly."

PRACTICAL PROBLEMS

1. $7,700 \div 25$ _____

2. $2,520 \div 45$ _____

3. $230,520 \div 85$ _____

4. $18,048 \div 32 \div 3$ _____

5. $[(475 - 85) \div (6 + 7)]$ _____

6. $2,730 \times [(375 \div 15) - (22 \div 2)]$ _____

7.   Find the prime factorization of 64.   _____

8. Find the prime factorization of 420.   _____

9. Find the average of 85, 83, 79, 95, 92, and 88.   _____

10. Find the average of 386, 455, 287, 820, 485, and 723.   _____

!11. Calculators can be used to find both the quotient and remainder in a
division problem. For example, enter the problem 74 ÷ 23 and press the
[=] key. The display should be 3.217391304 (or vary slightly depending
on the number of digits your calculator displays). Subtract the whole
number part of the quotient (3), and then multiply by the original divisor
(23). The display should be the remainder (in some cases, there may
be a small rounding error). Write a clear, concise explanation of why
this process works.   _____

12. How many shipping crates are needed if 22 plastic computer cases can
be placed in each crate and the total order to be shipped is 308 cases?   _____

13. A purchasing agent needed to order 12,240 O–rings for a production
run. The O–rings must be ordered by the gross (144). How many gross
should be ordered?   _____

14. A driver who was road testing tires traveled 348 miles in 6 hours. What
was her average speed in miles per hour? (Hint: When you see a unit
such as "miles per hour," remember to divide the number whose unit is
miles by the number whose unit is hours.)   _____

15. A manufacturing company has a 350-gallon tank of fuel oil. If the fuel is
burned at the rate of 25 gallons per day, how many days can the plant
operate on one full tank?   _____

16. A cubic foot is a measurement of volume; it is equal to the amount of
space contained in a cube which is 1 foot on each side. Loose foam
packing material is purchased in containers of 25 cubic feet. The current
inventory shows 14 containers of the material in stock. If 2 cubic feet of
material are used to pack each product carton, how many cartons can
be packed with the current inventory?   _____

17. The multiview drawing of a block below has overall dimensions length (L) = 4 inches, width (W) = 4 inches, and height (H) = 2 inches.

   a. If the views are to be positioned on the paper so that the horizontal spaces between borders and views are to be equal, find the spacing distance.

   _____

   b. If the views are to be positioned so that all spaces are equal vertically, what is that spacing distance?

   _____

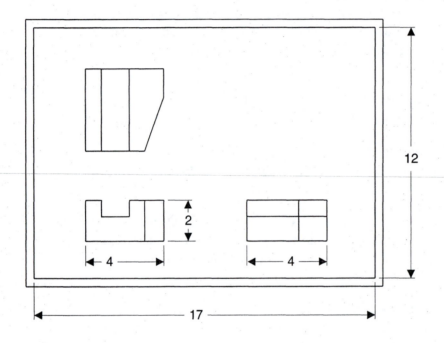

18. The quarterly records for two salespersons are shown. What is each salesperson's average quarterly sales? (The *average* value for a group of values is found by adding the values, and dividing by the number of values.)

Stevens: _____

Wilson: _____

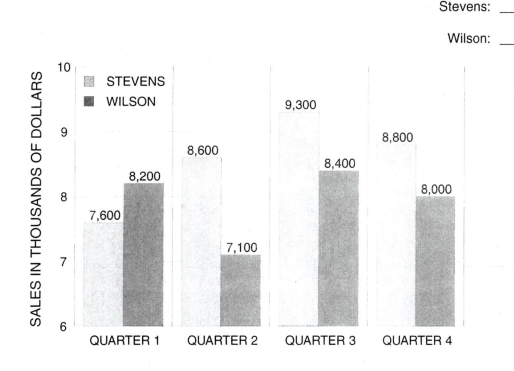

19. A manufacturing class is studying a unit on entrepreneurship. They have determined that they need $870 to manufacture and package their product.

   a. If they are selling shares of stock at $5 each, how many shares do they need to sell to finance their operation?   _____

   b. If the shares are sold at $15 each, how many shares must they sell?   _____

20. The size of a house is often stated in square feet. One square foot is an area equal to a piece 1 foot by 1 foot. The cost of a new house is estimated by the contractor to be $86,480. The house contains 1,880 square feet of living space. What is the cost per square foot for the living area?   _____

21. A large manufacturing plant has decided to purchase bicycles for workers to use as transportation between the manufacturing areas, offices, and tool crib due to the long distances. If they have estimated that they will need one bicycle for every 12 workers, and the maximum number of workers on any shift is 648, how many bicycles should they purchase? _____

22. Density is one characteristic of a material and is defined as mass divided by volume. If 8 cm^3 of a metal has a mass of 72 g, find its density. Include the units of grams per cubic centimeter (g/cm^3) in your answer. _____

23. Pressure is defined as force divided by area. If a force of 1,940 newtons is applied over an area of 20 square meters, what is the pressure in newtons per square meter (N/m^2)? _____

24. If a force of 3,810 pounds is applied to a post with an area of 127 square inches, what is the pressure in pounds per square inch (lb/in^2)? _____

25. The production times required for the first six air compressors on a new assembly line were recorded as 47 minutes, 55 minutes, 48 minutes, 41 minutes, 52 minutes, and 45 minutes. What is the average production time? _____

26. One characteristic of a spring is the "spring constant" (k), which is calculated using the formula $k = \dfrac{F}{\Delta L}$ where F is the force applied to stretch or compress the spring, and ΔL is the change in length when the force is applied. For a spring which compresses 5 inches when a load of 85 pounds is applied, what is the spring constant? _____

27. A supplier needs to assemble washer repair kits for a plumbing company. Each repair kit needs to have 6 small cone washers, 7 large cone washers, 12 medium flat washers, 8 large flat washers, and 5 small flat washers. The quantities currently in stock are 380 small cone washers, 412 large cone washers, 890 small flat washers, 680 medium flat washers, and 758 large flat washers. What is the maximum number of packages which can be assembled using the current inventory? _____

Unit 5 Review and Combined Operations on Whole Numbers

PLANNING A STRATEGY FOR MULTIPLE STEP PROBLEMS

The key to success in multiple step problems is to *plan your work, then work your plan.* Also, it is important when solving a problem to identify which information is relevant to a problem; often, the problem description contains numbers which are not needed in the calculations.

There are many ways to represent and work through a practical problem. Two methods will be presented here as examples. The first method is often used in engineering schools, and involves breaking the problem down into separate steps, with the results of each step being clearly labeled for use in further steps. The use of sketches to clarify the problem may be helpful. The second process is often used in mathematics classes and involves expressing the entire problem in math expressions to be simplified or equations to be solved. Both examples are illustrated below, using the same problem. In each case, the problem was worked using cents to avoid having decimals in the numerical values; you need to decide for a given problem what type of numerical representation is appropriate.

You may prefer one of these methods, or may want to develop your own system. A common myth is that "good math students don't have to write the steps." In reality, those who are successful in math, technology, and engineering typically write *more steps* and *clearly label* the problem parts.

Example: A buyer is monitoring the material use in the packaging department to determine the average cost of packaging for a set of wrenches, including waste. In the past three months, they used 2,736 perforated 7"x12" cardboard sheets at 35 cents each, 1,776 feet of shrinkwrap at 5 cents per foot, and 2,520 labels at 18 cents each. What is the total cost? If 2,500 wrench sets were packaged, what is the average cost per set for packaging materials?

Solution 1 Identify what you need to know to get the answer. Develop steps to get the intermediate information and clearly identify each value.

1. Determine how to represent the problem. In this problem, the values will be represented as cents to avoid including decimal representations.

2. Find the total cost of each type of material. (Multiply quantity of each by its unit cost.)

cardboard: 2,736 x 35 cents = 95,760 cents

shrinkwrap: 1,776 feet x 5 cents/foot = 8,880 cents

labels: 2,520 x 18 cents = 45,360 cents

3.   Add values from step 2 to find the total cost of materials.

cardboard + shrinkwrap + labels = total cost

95,760 + 8,880 + 45,360 = 150,000 cents

4.   Divide total cost by the number of packages to find the cost per package.

150,000 cents / 2,500 packages = 60 cents/package

Solution 2 Write one expression for the costs per sheet, including all needed values. In this example, the expression is written in words and then in math form for illustration. You may write the word form if it is helpful on more complicated problems, but it is not necessary. Hint: When simplifying expressions, use an equal sign to indicate that the expression following the equal sign has the same value (but a simpler appearance) as the one before the equal sign. An equal sign is not a "divider" between any numbers you want to keep separate: *it should be used only when the values on either side are equal in value!*

$$\frac{(boards \times unit\ cost) + (wrap \times unit\ cost) + (labels \times unit\ cost)}{number\ of\ packages}$$

$$\frac{(2,736 \times 35\ cents) + (1,776 \times 5\ cents) + (2,520 \times 18\ cents)}{2,500\ packages}$$

$$= \frac{(95,760 + 8,880 + 45,360)\ cents}{2,500\ pkg.} = \frac{150,000\ cents}{2,500\ pkg.} = \frac{60\ cents}{pkg.}$$

PRACTICAL PROBLEMS

1.   2,442 × 368   _____

2.   12,480 − [6,345 − (1,212 + 3,925)]   _____

3.   1,026 ÷ [43 + (42 ÷ 3)]   _____

4.   (322 + 475 + 286 + 325 + 332) ÷ 5   _____

5.   {[(20,860 ÷ 20) + 7] ÷ (3 x 5)}   _____

6. Find the prime factorization of 660. _____

7. Find the prime factorization of 175. _____

8. Find the average of 295, 312, 480, and 1,013. _____

9. 546 x {[(462 ÷ 11) + (37 x 8)] ÷ 2} _____

10. 528 ÷ {[44 − (4 x 8)] x (87 − 43)} _____

11. On a job, materials cost $385. Two workers, earning $8 per hour, worked 6 hours each. One worker, earning $7 per hour, worked 5 hours. What is the total cost of the job, including materials and labor? _____

12. The inventory of computer cable shows the following:

Thick Ethernet	725 feet
Thin Ethernet	1,225 feet
Twisted pair	1,880 feet

    If the company wants to have a stock of 3,000 feet of each cable, how much of each type should they purchase?

    Thick Ethernet _____

    Thin Ethernet _____

    Twisted Pair _____

13. A chemical plant had a full 250-liter tank of chemical A on July 15. During the next week, withdrawals of 54 liters, 32 liters, 65 liters, and 20 liters were made.

    a. How much chemical was taken from the tank that week? _____

    b. How much remained in the tank after the withdrawals? _____

14. An automotive service department sold 36 batteries, 24 tires, and 112 quarts of motor oil during a week. If they are required to charge a disposal fee of 85 cents for each battery, 75 cents for each tire, and 2 cents for each quart of oil, how much did they collect in disposal fees that week? Write your answer in cents. _____

15.   A lumber yard had 115 bundles of shingles in stock on September 1. On September 3, they received a partial shipment of 184 bundles from the factory. The remainder of the order, 64 bundles, arrived the following Friday.

  a.   How many bundles did they have in stock after the shipments?   _____

  b.   If it takes 3 bundles of shingles for 1 square (100 square feet), how many squares of shingles do they have?   _____

16.   A company is closing out its inventory of one style of cabinet. If they have 22 cabinets in stock, and paid $1,232 total for all the cabinets, how much should they charge per cabinet if they just want to "break even," or get their cost back?   _____

17.   A student at a technical school pays tuition of $52 per semester hour.

  a.   If he is enrolling for 16 semester hours, what will his tuition charge be?   _____

  b.   If books are estimated to cost $215 and general fees are $55, what are his total estimated costs?   _____

  c.   What is the tuition cost for a 4-hour class?   _____

18.   A manufacturer's representative travels an average of 25,200 miles per year by automobile. He is planning to buy a new car and is comparing Car A, priced at $14,000 with a fuel economy rating of 28 miles per gallon, to Car B, priced at $16,800 with a fuel economy rating of 36 mpg. Over a 2-year period, what are the expected savings on fuel costs for car B over car A? Use a gasoline price of $1 per gallon.   _____

19.   To build a set of shelves, 132' of 1" steel tubing are needed. There are 9 pieces, each 12' long in inventory.

  a.   Is the current inventory sufficient?   _____

  b.   If yes, how much will be left after the shelves are built? If no, how many more 12' pieces must be bought?   _____

20. An engineering technician compared purchasing a computer system as individual components rather than as a complete system. The prices of the components she selected were CPU and keyboard $1,095, monitor $459, mouse $38, 3½" floppy disk drive $62, and multimedia upgrade kit $339. Another store has a sale on a complete system with the same components priced at $2,099. Which is the better buy? _____

21. An automotive body shop charged a customer $575 to repair hail damage to a car. If the material costs were $103 for a windshield, $36 for a chrome trim strip, and $28 for paint, and the labor required was 5 hours at an hourly rate of $42, what was the profit on the job? _____

22. A company had 4 employees each earning $8 per hour, 2 employees each earning $9 per hour, and 2 trainees earning $6 per hour. If each worker worked for 36 hours, what was the total payroll? _____

23. How many cartons of #10–24 screws must be ordered if each carton contains 20 boxes with 144 screws per box and 129,600 are needed for a production run? _____

24. A subassembly for a printer has five parts, weighing 22 grams, 48 grams, 42 grams, 85 grams, and 36 grams.

    a. What is the total weight of the subassembly? _____

    b. If the pick–and–place robot that is currently installed on the production line has a maximum load rating of 300 grams, can the current robot be used to move the subassembly? _____

25. A plastics injection molding company recorded the number of occurrences of an incomplete fill in the molds, which resulted in rejected parts. The values for July – December are shown in the chart below. How many total parts were rejected for that reason during the six–month period? _____

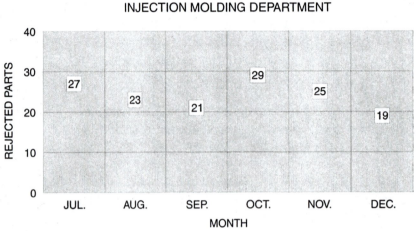

INCOMPLETE FILL REJECTS
INJECTION MOLDING DEPARTMENT

26. A car gasoline tank has a capacity of 15 gallons. If the typical gas mileage for the model is 36 miles per gallon, what distance should the driver be able to drive on one tank of gas? _____

27. A multiview drawing of a block is shown.

   a. If the views are to be positioned on the paper so that the horizontal spaces between borders and views are to be equal, find the spacing distance. _____

   b. If the views are to be positioned so that all spaces are equal vertically, what is that spacing distance? _____

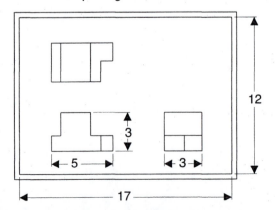

28. Ohm's Law states that the mathematical relationship between voltage (E), amperage (I), and resistance (R) in a circuit can be expressed as $I = E \div R$. If a circuit has a voltage of 216 volts and a resistance of 54 ohms, what is the current? _____

29. A construction company used the following materials on a job: concrete and steel $860, framing lumber $528, hardware $88, roofing materials $446, wiring and electrical supplies $212, drywall supplies $134, and paint supplies $96. If three workers each worked for 28 hours on the job, and each earns $14 per hour, what is the total cost of the job? _____

30. The quarterly records for two salespersons are shown.

   a. What is each salesperson's average quarterly sales?   Davidson _____

                                                                            Thomas _____

   b. What is the difference in the averages?   _____

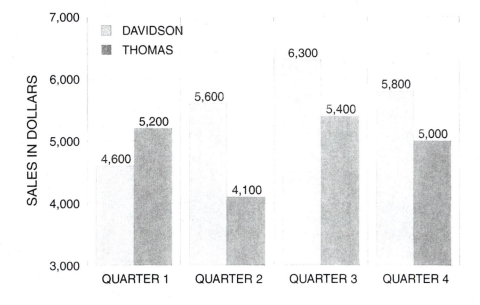

Common Fractions

SECTION
3

BASIC PRINCIPLES OF COMMON FRACTIONS

Common fractions are another type of number, often referred to as *rational numbers* in mathematics books. They are one way to express quantities which are not complete units. They also allow measurements to be expressed in smaller increments than whole numbers. Common fractions are used extensively in the construction, plumbing, automotive, and fastener industries.

A common fraction has three parts: a fraction bar (line between the two numbers), the numerator (number above the fraction bar), and the denominator (number below the fraction bar). The *denominator* of a fraction indicates the number of equal parts into which a whole unit has been divided. The *numerator* of a fraction indicates the number of those pieces used in the measurement. (In many typewritten or word-processed documents, the fraction may be written with a / between the numerator and denominator, as in ¼.)

Like fractions have the same denominator. *Unlike fractions* have different denominators.

Common fractions are often represented graphically as shown below. In the first illustration (¼ or "one fourth"), the unit is divided into 4 equal pieces and 1 of them is used in the fraction. In the second illustration (⅜ or "three eighths"), the unit is divided into 8 equal pieces and 3 of them are used in the fraction.

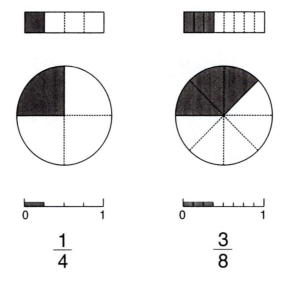

$$\frac{1}{4} \qquad\qquad \frac{3}{8}$$

The graphic form in the bottom illustration is used often in Industrial Technology. In construction related industries, inches are commonly divided into 16 equal parts (as seen on rules or steel tapes). In manufacturing, inches are often divided into 10 equal parts, or into multiples of 10. Although they can be expressed as fractions, they are more commonly written in decimal form.

Improper Fractions

An improper fraction has a larger number as a numerator than as a denominator. For example, ⁷⁄₄ and ²³⁄₈ are improper fractions. Although improper fractions are often accepted in algebra problems, they are usually converted to mixed number form for practical applications.

Mixed Numbers

A mixed number is a way of expressing a value greater than 1. It contains a whole number part and a fractional part, as in 4⅕.

WRITING EQUIVALENT FRACTIONS

A fraction can be written in many different forms, all of which have the same value. For example, ½ is equivalent to ¾ or ⅜ as shown below.

EQUIVALENT FRACTIONS

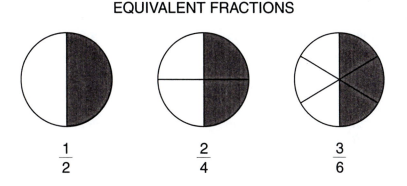

$$\frac{1}{2} \qquad \frac{2}{4} \qquad \frac{3}{6}$$

Raising Fractions to Higher Terms

To raise a fraction to higher terms, multiply the numerator and denominator by the same number. Since multiplying both numerator and denominator of a fraction by the same value is the same as multiplying by 1, the value of the fraction is not changed, although its form or appearance is different.

Example: Write ⅖ as an equivalent fraction with a denominator of 10. Since 5 must be multiplied by 2 to get the desired denominator of 10, multiply both numerator and denominator of the fraction by 2.

$$\frac{2}{5} \times \frac{2}{2} = \frac{4}{10}$$

Reducing Fractions to Lowest Terms

Fractions should always be left in lowest terms when calculations are finished. To reduce a fraction to lowest terms, determine a number which is a factor of both the numerator and denominator. This process is also referred to as "cancelling a common factor."

Example: Write 9/12 in lowest terms.

$$\frac{9}{12} \div \frac{3}{3} = \frac{3}{4}$$

Example: Write 20/24 in lowest terms.

$$\frac{20}{24} \div \frac{4}{4} = \frac{5}{6}$$

If you did not cancel out the greatest common factor (the largest factor that will divide into both numbers) the first time, and common factors remain, repeat the process until the numerator and denominator do not have any more common factors.

Example: Reduce 24/36 to lowest terms.

$$\frac{24}{36} \div \frac{3}{3} = \frac{8}{12} \div \frac{2}{2} = \frac{4}{6} \div \frac{2}{2} = \frac{2}{3}$$

An alternate method is to write the numerator and denominator of the fraction in prime factor form, and then cancel out any common factors.

Example: Reduce 24/36 to lowest terms.

$$\frac{24}{36} = \frac{2 \times 2 \times 2 \times 3}{2 \times 2 \times 3 \times 3} = \frac{2}{3}$$

Writing a Whole Number as a Fraction

A whole number can be written as a fraction by writing the number in the numerator and 1 in the denominator.

Example: $3 = \frac{3}{1}$ $8 = \frac{8}{1}$

Writing an Improper Fraction as a Mixed Number

To convert an improper fraction to a mixed number, divide the denominator into the numerator. Write the quotient as a whole number and place the remainder (if any) over the original denominator.

$$\frac{21}{4} = 5\frac{1}{4}$$

$$\begin{array}{r} 5 \text{ R } 1 \\ 4\,)\overline{21} \\ \underline{20} \\ 1 \end{array}$$

Writing a Mixed Number as an Improper Fraction

When a measurement or value is stated as a mixed number and must be multiplied or divided by other values, the mixed number can be converted to an improper fraction. To convert a mixed number into improper fraction form, multiply the denominator by the whole number and add the numerator. The resulting number is written as the numerator of the improper fraction, with the original denominator kept. In the example below, 5 is multiplied by 3, and the 1 is added to get 16. The 5 is kept as the denominator.

$$3\frac{1}{5} = \frac{16}{5} \qquad 3 \times 5 + 1 = 16$$

FINDING COMMON DENOMINATORS

A common denominator is a number that is a multiple of both numbers. In other words, each denominator should divide evenly (with no remainder) into the common denominator. For denominators of 6 and 18, the smallest common denominator is 18, although 36 could also be used. For denominators of 3 and 4, the smallest common denominator is 12, although 24 and 36 could also be used. If one denominator will divide evenly into the other, the larger number is the smallest common denominator.

Example: Write ⅔ and ⅘ as fractions with a common denominator. Since the lowest common denominator for 3 and 5 is 15, both fractions need to be written with a denominator of 15.

$$\frac{2}{3} \frac{\times 5}{\times 5} = \frac{10}{15} \qquad \frac{4}{5} \frac{\times 3}{\times 3} = \frac{12}{15}$$

It is usually easier to perform the math operations if the common denominator you choose is the *lowest common denominator* (LCD).

 CALCULATOR USE

Many scientific calculators allow fractions and mixed numbers to be entered in that form. The key which is used to indicate fractions is usually marked **a^b/c**. Some calculators also have a key marked **^{d}/c** or similar. The **a^b/c** key is normally used between the parts of a fraction or mixed number, for example between the 3, 1 and 5 in the mixed number 3⅕. After the fraction has been entered, the **^{d}/c** key can be pressed to display the fraction as an improper fraction. Your calculator may use a different sequence of keystrokes or key symbols. Also, many calculators have limits on the number of digits which can be entered as a numerator or denominator. On some calculators, when a fraction is entered and the equal key [=] is pressed, the fraction will be displayed in reduced form. When the **a^b/c** key is pressed repeatedly, the number displayed will alternate between the reduced fraction form and the decimal form. Try a fraction such as ¹²⁵/₆₀₀, which is equal to ⁵/₂₄ or 0.20833, to see how you calculator responds.

If your calculator has fraction capability, be sure that you are familiar with the way your calculator handles fractions.

PRACTICAL PROBLEMS

For problems 1–3 write equivalent fractions by filling in the missing numerator or denominator.

1. $\dfrac{2}{3} = \dfrac{}{6} = \dfrac{12}{} = \dfrac{}{21}$

2. $\dfrac{3}{8} = \dfrac{}{16} = \dfrac{12}{} = \dfrac{}{64}$

3. $\dfrac{3}{5} = \dfrac{}{10} = \dfrac{15}{} = \dfrac{}{40}$

For problems 4–8, reduce each fraction to lowest terms.

4. $\dfrac{22}{32}$ _____

5. $\dfrac{20}{64}$ _____

6. $\dfrac{42}{64}$ _____

7. $\dfrac{36}{48}$ _____

8. $\dfrac{56}{128}$ _____

9. Express each whole number as a fraction.

    3 _____   12 _____   8 _____   15 _____

For problems 10–15, find the least common denominator for the fractions. Then write each fraction as an equivalent fraction with that denominator.

10. $\dfrac{2}{3}$ $\dfrac{5}{8}$ _____

11. $\dfrac{1}{8}$ $\dfrac{8}{16}$ _____

12. $\dfrac{5}{8}$ $\dfrac{11}{64}$ _____

13. $\dfrac{2}{3}$ $\dfrac{4}{5}$ _____

14. $\dfrac{15}{16}$ $\dfrac{5}{12}$ _____

15. $\dfrac{11}{64}$ $\dfrac{13}{32}$ _____

For problems 16–20, convert each improper fraction to whole number or mixed number form.

16. $\dfrac{128}{32}$ _____

17. $\dfrac{55}{8}$ _____

18. $\dfrac{147}{8}$ _____

19. $\dfrac{243}{16}$ _____

20. $\dfrac{67}{12}$ _____

For problems 21–25, convert each mixed number to improper fraction form.

21. $7\dfrac{2}{5}$ _____

22. $10\dfrac{2}{3}$ _____

23. $21\dfrac{3}{8}$ _____

24. $4\dfrac{3}{16}$ _____

25. $21\dfrac{7}{8}$ _____

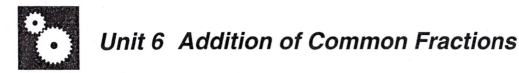

Unit 6 Addition of Common Fractions

BASIC PRINCIPLES OF ADDITION OF COMMON FRACTIONS

When fractions are added, their denominators must be the same, meaning that the same size part is described by all of the fractions. For example, they are all "fourths" or "sixteenths," etc. The sum of the fractions is obtained by adding the numerators and keeping the common denominator.

$$\frac{3}{16} + \frac{7}{16} = \frac{10}{16} = \frac{5}{8}$$

The sum is illustrated below using a ruler with each inch divided into sixteen equal parts (¹⁄₁₆ inch each).

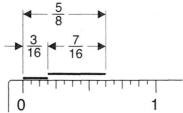

Another common rule used in Industrial Technology is referred to as a civil engineer's scale, with inch units divided into 10, 20, 30, 40, 50, or 60 pieces each. In the illustration below, the sum ³⁄₁₀ + ⁵⁄₁₀ = ⁸⁄₁₀ is illustrated using the 10 scale.

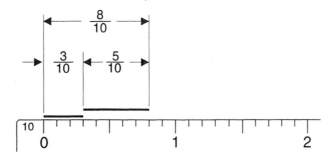

If the denominators of the fractions are different, each fraction must be expressed as an equivalent fraction with a common denominator. The addition is then completed using the steps described above. In the example below, the second fraction ¼ is changed into ⁴⁄₁₆ by multiplying both its numerator and denominator by 4. The fractions then have common denominators and can be combined to find the total number of "sixteenths."

$$\frac{3}{16} + \frac{1}{4} = \frac{3}{16} + \frac{1(4)}{4(4)} = \frac{3}{16} + \frac{4}{16} = \frac{7}{16}$$

Remember to express the resulting fraction in lowest terms or reduced form.

Addition of Mixed Numbers

Mixed numbers can be added in that form by adding the whole number parts and adding the fractional parts. If the resulting fraction is improper, it can be rewritten as a mixed number and added to the whole number.

Example:
$$3\frac{2}{3} + 2\frac{5}{8} = (3+2) + (\frac{2}{3} + \frac{5}{8})$$
$$= 5 + (\frac{16}{24} + \frac{15}{24})$$
$$= 5 + 1\frac{7}{24}$$
$$= 6\frac{7}{24}$$

As an alternate process, the mixed number can be written as improper fractions and added. The sum can then be converted to a mixed number.

 CALCULATOR USE

If your calculator has fraction capability, it can be used to perform addition of fractions and mixed numbers. When entering mixed numbers on some calculators, you may need to enter a plus sign [+] between the whole number and fraction parts.

PRACTICAL PROBLEMS

1. $\dfrac{4}{27} + \dfrac{5}{27}$ _____

2. $\dfrac{11}{32} + \dfrac{13}{32}$ _____

3. $\dfrac{3}{16} + \dfrac{15}{64}$ _____

4. $\dfrac{7}{16} + \dfrac{9}{40}$ _____

5. $\dfrac{11}{15} + \dfrac{47}{50}$ _____

6. $3\dfrac{1}{5} + 7\dfrac{2}{5}$ _____

7. $23\dfrac{7}{16} + 8\dfrac{3}{16}$ _____

8. $12\dfrac{9}{32} + 9\dfrac{11}{16}$ _____

9. $9\dfrac{27}{64} + 3\dfrac{7}{8}$ _____

10. $\dfrac{33}{24} + \dfrac{11}{32}$ _____

11. A floor has 2 layers of plywood, one ¾" thick and the other ⅝" thick. What is the total thickness of the two layers of plywood? _____

12. If the floor in problem 11 is covered with a layer of vinyl tile which is ⅛" thick, what is the total thickness of the plywood and tile? _____

13. A roof has 2 × 6 rafters (5½" thick), a layer of ¾" plywood decking, felt ¹⁄₁₆" thick, and a layer of shingles ³⁄₁₆" thick. What is the total thickness of all the roof components? _____

14. A motor and support block is shown below. What is the total height from the bottom of the block to the center of the motor shaft? _____

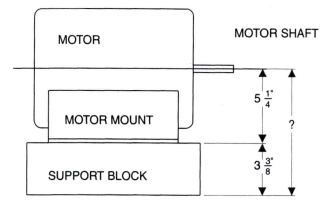

15. An automotive repair technician worked for 2¼ hours on Tuesday and 3¼ hours on Wednesday repairing a transmission.

   a. Find his total time.

   _____

   b. If the standards book states that the repair should require 6¼ hours, was the technician's actual time over or under the standard?

   _____

   c. By how much?

   _____

16. A support bracket is formed by bending a piece of sheetmetal. Neglecting the bend allowance, what is the total length of the sheetmetal piece before bending?

   _____

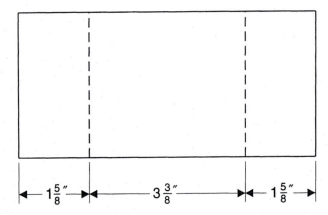
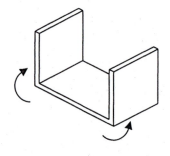

17. What is the total thickness of a brick veneer wall with a 3½" stud, ½" sheetrock on the interior, ½" exterior gypsum sheathing, a ¾" air space, and 4¼" brick?

   _____

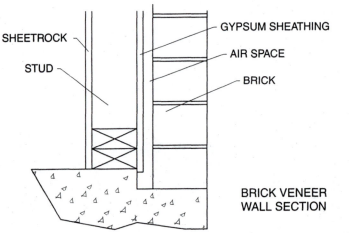

SHEETROCK

STUD

GYPSUM SHEATHING

AIR SPACE

BRICK

BRICK VENEER
WALL SECTION

18. For the sectional assembly view shown;

    a. find the height of the assembly using the nominal dimensions.   _____

    b. find the height of the assembly if each part has the maximum dimension allowed.   _____

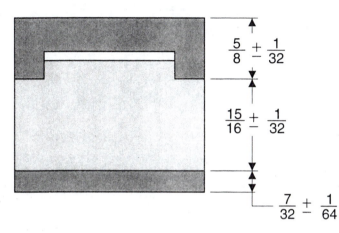

19. A pattern for an open-top rectangular box is shown below. What are the minimum dimensions for a piece of sheet aluminum from which the box pattern can be cut?   _____

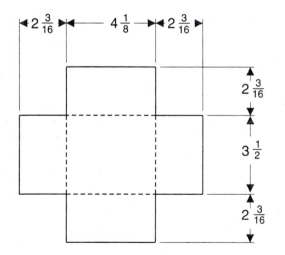

20. A $\frac{7}{16}$" diameter rod is covered with an insulating coating $\frac{5}{32}$" thick. What is the diameter including the covering?   _____

21. Five resistors, with individual resistances of ⅝ ohm, ⅓ ohm, ¼ ohm, ½ ohm, and ⅖ ohm are connected in series. For resistances connected in series, the total resistance is the sum of the individual resistances. Find the total amount of resistance.

_____

22. What is the minimum length of ⅞ inch diameter steel tubing from which the following pieces can be cut: 8⅛ inches, 15⅞ inches, 22¾ inches, and 6½ inches? Allow ¹⁄₁₆ inch for each cut.

_____

23. Find the total cost of 3½ pounds of 8d box nails at 78 cents per pound, 2¼ pounds of #4 finish nails at 68 cents per pound, and 1¾ pounds of #6 finish nails at 64 cents per pound. Find the total in cents and then write the answer as dollars and cents.

_____

24. For the angle plate shown, find the missing dimensions.

A = _____

B = _____

C = _____

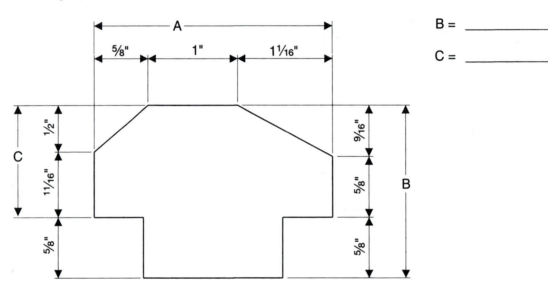

25. A lathe was used to turn a motor shaft. If the finished diameter was 1⁹⁄₁₆ inches and ⁵⁄₃₂ inch was removed from the diameter, what was the original diameter?

_____

Unit 7 Subtraction of Common Fractions

BASIC PRINCIPLES OF SUBTRACTION OF COMMON FRACTIONS

Subtraction of common fractions is similar to addition of fractions. For fractions which have the same denominator, subtract the smaller numerator from the larger numerator and write the difference as the numerator of the answer. The denominator is the same as the denominator in the fractions.

$$\frac{9}{12} - \frac{5}{16} = \frac{4}{16} = \frac{1}{4}$$

If the fractions have different denominators, rewrite each fraction as an equivalent fraction having a common denominator. Then follow the steps in the paragraph above.

$$\frac{5}{8} - \frac{1}{4} = \frac{5}{8} - \frac{1(2)}{4(2)} = \frac{5}{8} - \frac{2}{8} = \frac{3}{8}$$

Remember to write the resulting fraction in lowest terms or reduced form.

Subtraction of Mixed Numbers

Mixed numbers can be subtracted in that form or as improper fractions. To subtract mixed numbers in that form, subtract the fraction parts first, regrouping from the whole number part as needed. Then subtract the whole numbers.

Example: $$3\frac{1}{4} - 1\frac{5}{8} = 2\frac{5}{4} - 1\frac{5}{8} = (2-1) + (\frac{10}{8} - \frac{5}{8}) = 1\frac{5}{8}$$

To subtract as improper fractions, write each in common denominator form and subtract.

 ## CALCULATOR USE

If your calculator has fraction capability, it can be used to perform subtraction of fractions and mixed numbers. When entering mixed numbers, remember to enter a plus sign [+] between the whole number part and the fraction part, if needed for your calculator.

PRACTICAL PROBLEMS

1. $\dfrac{7}{16} - \dfrac{3}{16}$ _____

2. $\dfrac{15}{32} - \dfrac{11}{32}$ _____

3. $\dfrac{27}{64} - \dfrac{3}{16}$ _____

4. $\dfrac{7}{8} - \dfrac{11}{40}$ _____

5. $\dfrac{29}{32} - \dfrac{11}{25}$ _____

6. $4\dfrac{1}{8} - 2\dfrac{5}{8}$ _____

7. $25\dfrac{5}{16} - 9\dfrac{3}{16}$ _____

8. $24\dfrac{5}{32} - 11\dfrac{5}{16}$ _____

9. $9\dfrac{7}{16} - 3\dfrac{7}{8}$ _____

10. $\dfrac{33}{24} - \dfrac{19}{64}$ _____

11. a. On Monday, Davis Bearing Co. stock fell from 23⅝ to 21⅞. How much did it drop?   _____

      b. On Tuesday, it rose to 24⅛. What was the increase on Tuesday?   _____

12. What is the distance (D) between the centers of the holes in the block? _____

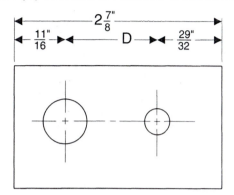

13. A bolt is 2½" long under the head. If it is used to join two pieces of steel which are 1⅜" thick and ½" thick, how much of the bolt will extend beyond the thickness of the steel plates? _____

14. A piece of pipe has an outside diameter of 1½". If the wall is ⅛" thick, what is the inside diameter? _____

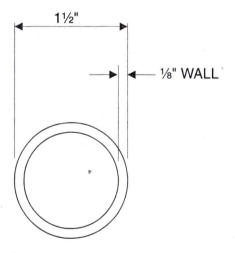

15. The term *tolerance* refers to the total variation (maximum size – minimum size) allowed for a dimension on a part. A measurement on a drawing is stated as 6³⁄₁₆" ± ¹⁄₁₆".

   a. What are the maximum and minimum dimensions?   maximum: _____

   minimum: _____

   b. What is the tolerance?   _____

16. A drawer is 15" wide inside, and is to be used to hold computer floppy disks. If the dividers are ½" thick and the spaces between dividers need to be 3¾", use repeated subtraction to determine how many spaces the drawer can hold. How much space (D) will remain on one side? _____

D = _____

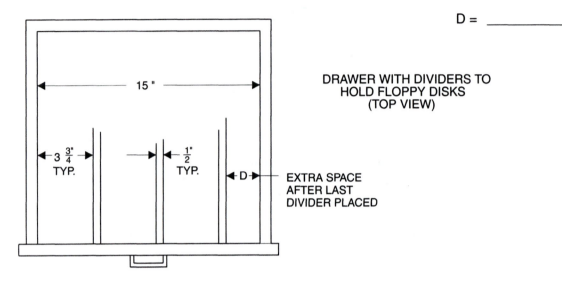

DRAWER WITH DIVIDERS TO
HOLD FLOPPY DISKS
(TOP VIEW)

15 "

$3\frac{3}{4}$"
TYP.

$\frac{1}{2}$"
TYP.

D

EXTRA SPACE
AFTER LAST
DIVIDER PLACED

17. A fixture for a part to be machined must hold the part so that its top surface is 4⅛" above the machine table. If the part is 2⁵⁄₁₆" thick, how thick must the fixture base be? _____

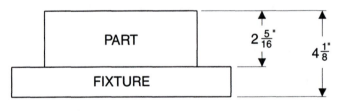

PART

FIXTURE

$2\frac{5}{16}$"

$4\frac{1}{8}$"

MACHINING FIXTURE FOR
MILLING OPERATION

18. A collar on a machine has a 6⅛" inside diameter. If the machine part on which it fits has an outside diameter of 6³⁄₆₄", how much larger is the collar diameter than the machine part diameter? _____

19. A steel rod has an outside diameter of 1⅛". If it has to be turned down to a diameter of ¹⁵⁄₁₆", how much must be removed from the diameter? _____

20. A drawing of a base plate is shown below. Find the missing dimensions A, B, and C.

A = _____

B = _____

C = _____

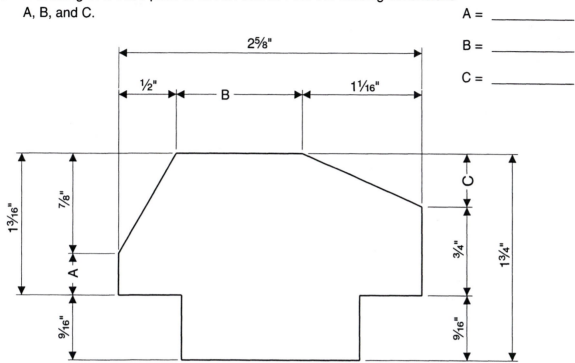

21. Two steel plates 3½ inches wide are positioned in a lap joint for welding. If the overlap of the two plates is ⅜ inch, what is the width (W) of the welded piece?

_____

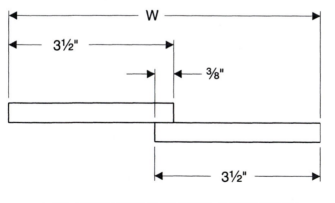

LAP JOINT FOR WELDING OPERATION

22. If the length of a precut stud is calculated so that the total height of the stud plus 2 top plates which are 1¹¹⁄₁₆ inches thick each is equal to 96 inches, what is the length of the precut stud? (Make a sketch of this problem, identifying all dimensions.)

_____

23. The rough opening specifications for a window installation require a ⅞-inch clearance on each side. If the actual window dimension is 31½ inches, what is the rough opening size?

_____

24. A tensile test specimen measured 2⁹⁄₁₆ inches after testing. If the original specimen length was 1⅞ inches, what was the amount of elongation (stretch) during the test?

_____

25. For the support base shown, find the missing dimensions.

A = _____

B = _____

C = _____

D = _____

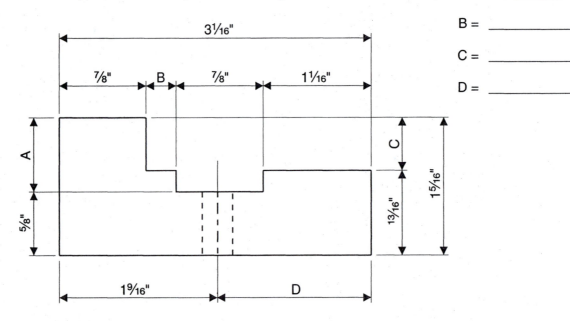

26. The thickness of a motor base was 1¹⁄₁₆ inches before milling the top surface. Two milling operations were performed, one rough cut to remove most of the material and one finish cut. If the thickness after milling was ⅞ inch and the finish cut removed ¹⁄₃₂" of material, how much material was removed during the rough milling cut?

_____

Unit 8 Multiplication of Common Fractions

BASIC PRINCIPLES OF MULTIPLICATION OF COMMON FRACTIONS

When fractions are multiplied, *it is not necessary to find a common denominator.* To multiply two fractions, multiply the numerators together and write the product as the numerator of the answer. Then multiply the denominators together and write the product as the denominator of the answer. The resulting fraction should be expressed in lowest terms.

$$\frac{5}{8} \times \frac{1}{2} = \frac{5}{16}$$

$$\frac{3}{8} \times \frac{4}{5} = \frac{12 \ (\div 2)}{40 \ (\div 2)} = \frac{6 \ (\div 2)}{20 \ (\div 2)} = \frac{3}{10}$$

When multiplying fractions, it is possible to reduce the problem before actually performing the multiplications. If a *common factor* (a number which will divide into both numbers) can be found for the numerator of one fraction and the denominator of the other, reduction can be done before the multiplication is performed. To illustrate that this simplification is mathematically correct, study the following example.

$$\frac{3}{8} \times \frac{4}{5} = \frac{3 \times 4}{8 \times 5} = \frac{12 \ (\div 4)}{40 \ (\div 4)} = \frac{3}{10} \qquad \frac{3}{\cancel{8}2} \times \frac{\cancel{4}^1}{5} = \frac{3}{10}$$

Example: Find the radius of a circle with a ⅞ inch diameter. Since the radius of a circle is ½ of the diameter, multiply ⅞ by ½.

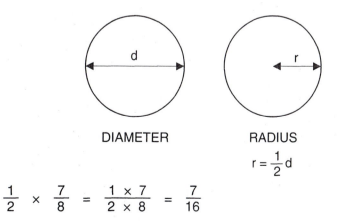

DIAMETER

RADIUS

$$r = \frac{1}{2}d$$

$$\frac{1}{2} \times \frac{7}{8} = \frac{1 \times 7}{2 \times 8} = \frac{7}{16}$$

Note: As described in Unit 3, the word "of" in practical problems often refers to multiplication, as in " ¾ of the 560 products were acceptable quality."

Multiplying Fractions by Whole Numbers

When multiplying a fraction by a whole number, write the whole number as a fraction with a denominator of 1.

Example: $\dfrac{7}{16} \times 3 = \dfrac{7 \times 3}{16 \times 1} = \dfrac{21}{16} = 1\dfrac{5}{16}$

Multiplying Mixed Numbers

When multiplying mixed numbers, write each mixed number in improper fraction form. Then multiply the two fractions.

Example: $2\dfrac{1}{4} \times 3\dfrac{1}{3} = \dfrac{9 \times 10}{4 \times 3} = \dfrac{30}{4} = \dfrac{15}{2} = 7\dfrac{1}{2}$

 CALCULATOR USE

If your calculator has fraction capability, fractions and mixed numbers can be multiplied. Remember to enter a plus sign [+] between the whole number and fraction parts of a mixed number if necessary. Be careful to enter the values in a way that ensures that the Order of Operations is followed. For example, to multiply 3½ by 2¼, you may need to use parentheses around the 3 and the ½ since their values need to be added before the multiplication is done.

PRACTICAL PROBLEMS

1. $\dfrac{9}{16} \times \dfrac{1}{3}$ _____

2. $\dfrac{15}{16} \times \dfrac{2}{3}$ _____

3. $\dfrac{9}{32} \times \dfrac{1}{2}$ _____

4. $2\dfrac{3}{8} \times \dfrac{1}{4}$ _____

5. $3\dfrac{1}{2} \times 2\dfrac{3}{4}$ _____

6. $(8)\ (3\frac{1}{2})\ +\ (5)\ (4\frac{1}{4})$

_____

7. $12\frac{1}{2}\ \times\ \frac{2}{3}$

_____

8. $1\frac{15}{16}\ \times\ \frac{3}{4}$

_____

9. $\frac{1}{4}(3\frac{5}{8}\ -\ 2\frac{1}{16})$

_____

10. $\frac{1}{3}[6\frac{7}{8}\ -\ (3)\ (1\frac{1}{8})]$

_____

11. For a set of 4 resistors, 5½ ohms each, connected in series, the total resistance is equal to the sum of the individual resistances.

  a. Write the formula for total resistance using multiplication notation.

_____

  b. What is the total resistance?

_____

12. Drawing boards are ¾ inch thick. If they are to be stacked in a storage cabinet, how tall will a stack of 32 boards be?

_____

13. A drafter worked 3¾ hours at $8 per hour. What are her total earnings?

_____

14. The weight of a bushing is 1¼ lb. What is the total weight of the bushings if an order of 125 parts is to be shipped? Write your answer as a mixed number.

_____

15. A carpenter needs to cut 8 pieces of 1x6, each 5½ inches long. Allowing ⅛" for each of the 7 cuts needed, what is the minimum length of material needed?

_____

16. A hole to be drawn on a working drawing is ¹³⁄₁₆ inch in diameter. In order to set the compass, a drafter needs to find the radius, which is ½ of the diameter. What is the radius?

_____

17. A 2 × 4 is to be ripped into 4 pieces to make furring strips. If the actual width of a 2 × 4 is 3½ inches, and the 3 saw cuts will use ⅛ inch each, what is the width (W) of each furring strip? Note that each strip is ¼ of the material remaining after the saw kerfs are subtracted.)

    _____

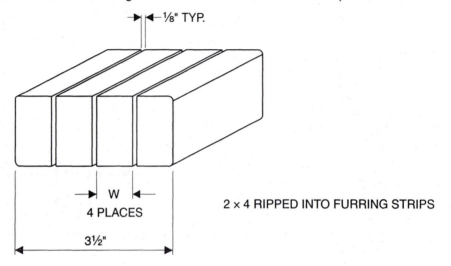

    ⟶|⟵ ⅛" TYP.

W
4 PLACES

3½"

2 × 4 RIPPED INTO FURRING STRIPS

18. The value π is often expressed as the fraction ²²⁄₇. The circumference of a circle is calculated by multiplying π by the diameter. Using ²²⁄₇ for π, what is the circumference of a circle with a diameter of 1¾ feet? Be sure to include the units of feet in your answer.

    _____

19. In a fastener factory, three machines produce ¼-20 × 1" machine screws. One machine produces ¾ pound per minute and the other two machines each produce 1¼ pounds per minute. If each machine runs for 45 minutes, how many pounds of screws are produced?

    _____

20. A plumbing job calls for 10 pieces of 1½ inch PVC pipe each 3½ feet long, and 6 pieces each 8¾ feet long. How many feet of 1½ inch PVC pipe are required?

    _____

21. In a machining operation on an aluminum part, the waste can be approximated as ⅙ of the weight of the part before machining.

   a. For a part weighing 2¼ pounds, what is the estimated amount of waste per part?

    _____

   b. For a production run of 640 parts, what is the estimated amount of waste?

    _____

22. What is the minimum length of aluminum channel needed to cut 14 pieces, each 1¾₁₆ inches long, if each cut uses ⅛ inch? (Draw a diagram as needed to determine the number of cuts.) _____

23. Using ²²⁄₇ as an approximate value for π, find the circumference of a pulley with a diameter of 14⅞ inches. _____

24. Using a system of movable pulleys, the amount of force needed to lift a weight can be reduced. For a system with 4 ropes supporting the weight to be lifted, the distance that the weight moves is ¼ the distance over which the force is applied. For an applied force distance of 3½ feet, what distance will the load be lifted? _____

25. The moment of a force (also known as torque) is the tendency of the force to cause a body on which it acts to rotate about an axis. Moment is the product of the magnitude of the force (F) and the distance (d), which is the perpendicular distance from the axis to the line of action of the force. For a force of ¾ pound and a distance (d) of 1¼ feet, what is the moment? Include the units of pound-feet in your answer. _____

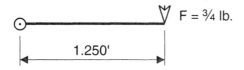

26. To calculate the sheetmetal bend allowance for soft metals, the thickness of the metal is multiplied by ⅓. The result is added to the inside dimension of the metal for each bend required.

    a. If the metal is ³⁄₃₂ inch thick, what is the bend allowance for each bend? _____

    b. How much should be added to the material length needed if 4 bends must be made? _____

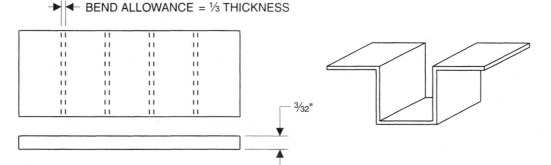

27. A model of a manufacturing cell is to be built to ⅛ actual size. If the dimensions of the space are 22½ feet long, 14 feet wide, and 9 feet high, what are the length, width, and height dimensions to be used for the model?

length _____

width _____

length _____

Unit 9 Division of Common Fractions

BASIC PRINCIPLES OF DIVISION OF COMMON FRACTIONS

Division of common fractions involves the use of the *reciprocal* of a fraction. To write the reciprocal of a fraction, change the positions of the numerator and denominator. For example, the reciprocal of ⅔ is 3/2. The reciprocal of ⅝ is 8/5.

Division of common fractions is a process similar to multiplication of common fractions. It does not require the use of a common denominator. To divide fractions, multiply the dividend (the number before the division sign) by the reciprocal of the divisor (the number after the division sign). To divide ¾ by ⅝, multiply ¾ by 8/5, and reduce the answer.

$$\frac{3}{4} \div \frac{5}{8} = \frac{3}{4} \times \frac{8}{5} = \frac{24}{20} = \frac{6}{5}$$

Dividing Mixed Numbers

To divide mixed numbers, change the mixed numbers to improper fractions and divide as described above.

 CALCULATOR USE

If your calculator has fraction capability, it can be used to perform division operations on fractions and mixed numbers. As for multiplication of mixed numbers, you may need to use parentheses around the mixed numbers so that the whole number parts and fraction parts are added before the mixed number value is divided. Pay special attention to the Order of Operations in problems which have multiple operations.

PRACTICAL PROBLEMS

1. $\dfrac{5}{8} \div \dfrac{3}{4}$ _____

2. $\dfrac{13}{16} \div \dfrac{1}{8}$ _____

3. $9\dfrac{7}{8} \div 2$ _____

4. $\dfrac{15}{16} \div \dfrac{2}{3}$ _____

5. $3\dfrac{1}{2} \div 2\dfrac{3}{4}$ _____

6. $(8\dfrac{5}{8} - 2\dfrac{1}{4}) \div 3$ _____

7. $11 - (2\dfrac{3}{4} + 1\dfrac{7}{8} + 1) \div 3$ _____

8. $\dfrac{2}{3} \times \dfrac{4}{7} \div \dfrac{1}{6}$ _____

9. $\dfrac{13}{32} \times \dfrac{1}{2} + \dfrac{1}{4} \div \dfrac{7}{16}$ _____

10. $\dfrac{5}{16} \div \dfrac{1}{4}$ _____

11. One way of evaluating bridge models is to calculate a failure factor by dividing the weight held by the weight of the bridge. If a bridge held 22 pounds and weighed 1¾₁₆ pounds, what is the failure factor?   _____

12. When a nut is placed on a #10-24 bolt, one revolution (turn) of the nut moves it ½₄" down the length of the bolt. How many turns will be required to move the nut ⅜" down the bolt?   _____

13. If a 90-inch piece of steel I-beam is to be cut into 7½" long pieces, how many can be cut? (Neglect any cutting loss.)   _____

14. The inside height of a cabinet is 21¾". If the space is to be divided into 5 equal-height shelf spaces for drafting paper storage and ¾" plywood is to be used for each of the 4 shelves needed, what is the vertical height of each shelf space?

_____

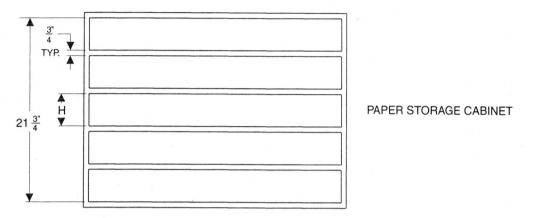

PAPER STORAGE CABINET

15. A drafter needs to lay out the top, front, and right side views of a block on a sheet of paper. For an 8½-inch by 11-inch sheet of paper, with a border ½ inch from each edge, find the horizontal spacing (X) between the views. Find the vertical spacing (Y).

X _____

Y _____

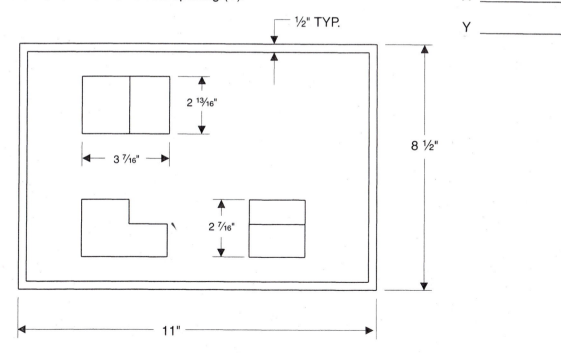

16. A circuit contains five identical resistors in series, with a total resistance of 8⅓ ohms. The total resistance is the sum of the individual resistors.

    a. Write a math expression using division which could be used to find the resistance value of each resistor.

    _____

    b. What is the resistance value of each resistor?

    _____

17. Ohm's Law states that amperage (I) = voltage (E) ÷ resistance (R). If the voltage in a circuit is 11 volts and the resistance is 2¾ ohms, find the amperage.

    _____

18. What is the hourly rate of pay if Ryan received $207 for 34½ hours of work?

    _____

19. The term "equally spaced" occurs often on drawings for manufactured parts. What is the distance between the centers of adjacent holes in the figure below?

    _____

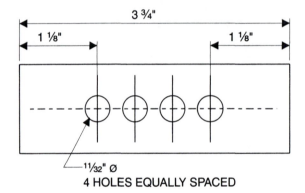

3 ¾"

1 ⅛"   1 ⅛"

11/32" Ø
4 HOLES EQUALLY SPACED

Unit 10 Review and Combined Operations on Common Fractions

PRACTICAL PROBLEMS

1. $\dfrac{7}{16} + 1\dfrac{1}{8}$ _____

2. $\dfrac{1}{32} \div \dfrac{1}{4}$ _____

3. $3\dfrac{1}{2} - 2\dfrac{7}{32}$ _____

4. Reduce ⁴⁸⁄₁₂₈ to lowest terms.   _____

5. Fill in the missing blanks:

   $\dfrac{5}{8} = \dfrac{}{16} = \dfrac{20}{} = \dfrac{}{56}$

6. $8\dfrac{5}{32} + 9\dfrac{11}{16}$ _____

7. $\dfrac{35}{24} - \dfrac{15}{64}$ _____

8. $12\dfrac{3}{8} \times 45$ _____

9. $1\dfrac{9}{32} - \left(2 \times \dfrac{5}{64}\right)$ _____

10. $\dfrac{1}{3}\left[10 - \left(2\dfrac{1}{4} + 1\dfrac{3}{8}\right)\right]$ _____

11. A block dimensioned using (a) sequential dimensioning and (b) datum dimensioning is shown below. Find the missing dimensions on the view dimensioned using datum lines.

A _____

B _____

C _____

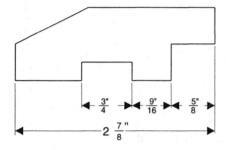
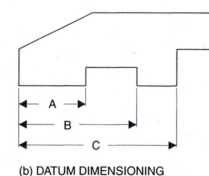

(a) SEQUENTIAL DIMENSIONING   (b) DATUM DIMENSIONING

12. A piece of card stock is 11 inches long. How many 2¾ inch strips can be cut from that length?

_____

13. Determine the length of each illustrated line segment from the drawing. Then perform the math operations indicated.

A _____

B _____

C _____

D _____

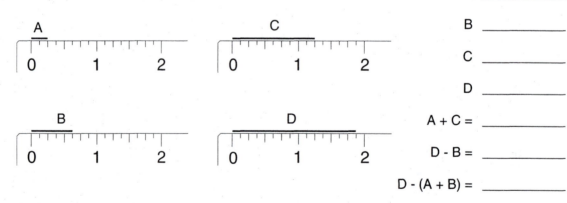

A + C = _____

D - B = _____

D - (A + B) = _____

14. A broken shaft on a machine is to be replaced, but the only available replacement part with the same diameter is longer than the original part. If the original part was 6⅞ inches long, and the replacement is 7⁵⁄₁₆ inches long, how much must be cut off of the new part?

_____

15. In printing, 1 point is ¹⁄₇₂ inch. How many points are there in ⅝ inch?

_____

16. Determine the length of each illustrated line segment from the drawing. Then perform the math operations indicated.

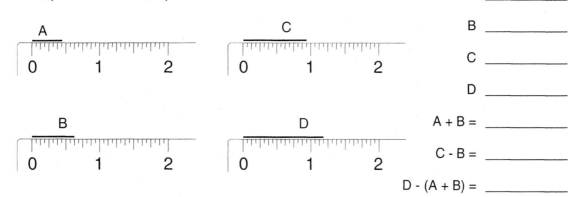

A _____

B _____

C _____

D _____

A + B = _____

C - B = _____

D - (A + B) = _____

17. The outside diameter of round tubing is 1⅛ inches. The inside diameter is ¾ inch. Make a sketch as needed. What is the wall thickness?

_____

18. A motor is positioned by placing shims under it. If the shims are 1½ inches, ⅞ inch, and 5⁄16 inch thick, what is the total thickness of all 3 shims?

_____

19. If a production run of lamp bases required 3 stamping machine operators working 32½ hours each and 2 finishing machine operators working 15¾ hours each, what is the total number of hours charged to the production run?

_____

20. A bolt is 1¾ inches long. It is to be used to fasten 2 pieces of aluminum plate which are ⅞ inch and ¼ inch thick. How much of the bolt will extend beyond the thickness of the plates?

_____

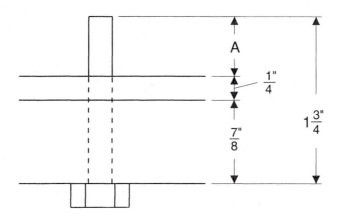

21. If ¾ inch plywood has 8 plies of equal thickness, what is the thickness of each ply?

_____

22. A piece of pipe has an outside diameter of 3½ inches and a wall thickness of ³⁄₁₆ inch. Make a sketch showing all relevant data and find the inside diameter.

_____

23. A steel rod has an outside diameter of ¾ inch. If it is to be turned down to ¹⁷⁄₃₂ inch, how much must be removed from the diameter?

_____

24. If the total shipping weight of an order is 38½ pounds and the order is packed in 7 identical boxes, what is the weight of each box?

_____

25. For the drawing shown, calculate the missing dimensions. If the dimension is less than 1 inch, write it as a fraction. If any dimension is over 1 inch, write it as a mixed number.

A _____

B _____

C _____

B = TOTAL LENGTH OF BLOCK

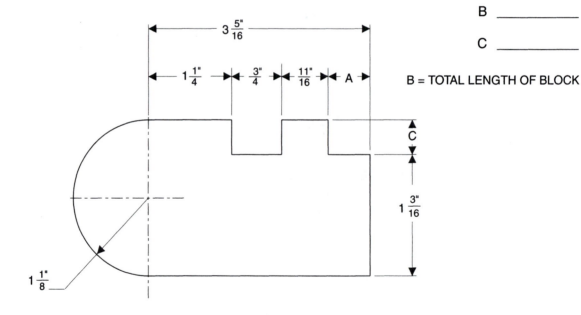

26. Use a ruler which is divided into 16 pieces per inch (¹⁄₁₆ inch each) and measure each of the lines to the nearest ¹⁄₁₆ inch. Place the zero (0) on the ruler exactly at one end of the line and read the measurement at the other end of the line.

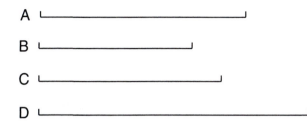

A _____

B _____

C _____

D _____

27. A trailer manufacturing company has five pieces of angle iron in a storage room, with length measurements of 8½ feet, 4¼ feet, 6¾ feet, 10 feet, and 8¾ feet. A worker is told to cut as many 1½ foot long pieces as possible from the stock. How many pieces can be cut from the five pieces? _____

28. The amount of bell wire in inventory is to be approximated by weight. The weight of the wire is listed in the catalog as 2¼ ounces per 100 feet. An empty spool of the same type used on the roll weighs 1¼ ounces. If the roll to be inventoried weighs 20⅜ ounces, what is the approximate amount of wire on the spool in feet? _____

29. For the drawing shown, find the missing dimensions.

A _____

B _____

C _____

D _____

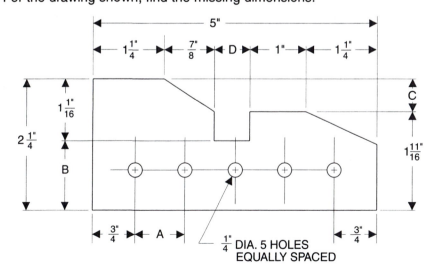

1/4" DIA. 5 HOLES EQUALLY SPACED

30. Find the inside dimension of square steel tubing with an outside dimension of 2³⁄₁₆ inches if the wall thickness is ⁵⁄₃₂ inch. _____

Decimal Fractions

SECTION

4

Decimal fractions are another type of number used to represent quantities which are not whole units. Mathematical operations on decimal fractions do not require the use of a common denominator. Decimal fractions allow operations on fractional values using math operations similar to those on whole numbers, along with handling the decimal point correctly.

DECIMAL PLACES

Decimal fractions are similar to whole numbers in that each place has a value based on its position. The place values of the first four decimal places in a decimal fraction are illustrated below.

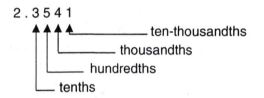

Decimal fractions are used frequently in industry, especially in manufacturing. Examples include dimensioning drawings in decimal form, reading micrometer measurements, temperature measurements, and money.

You should know how to read the names of decimal fractions orally. Read the name of the number to the right of the decimal as though it were a whole number and then add the name of the last decimal place. For example, 0.387 is read as "three hundred eighty seven thousandths."

Money

The U.S. currency system is a decimal system based on units called dollars.

Example: $ 2 1 . 8 5

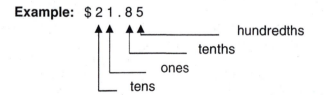

Using dollars as units, a dime equals ⅒ or 0.1 of a dollar, while a cent equals ⅟₁₀₀ or 0.01 of a dollar. A value expressed in cents can be written using dollar units by moving the decimal point 2 places to the left and changing the cents sign to a dollar sign.

Examples: 65¢ = $0.65

388¢ = $3.88

Unit 11 Significant Digits, Rounding, and Scientific Notation

BASIC PRINCIPLES FOR ROUNDING DECIMAL FRACTIONS

There are several ways to specify the level of accuracy to which a decimal fraction is to be expressed. This is often necessary because of the use of calculators which can easily calculate and display many digits. A number can be rounded to a specified level of accuracy (number of significant digits) or to a specified level of precision (number of decimal places).

ROUNDING TO DECIMAL PLACES (PRECISION)

In many industrial applications, especially machining and production, the number of decimal places is specified according to the physical characteristics of the data. For example, in product dimensions, the number of decimal places used may be related to the manufacturing process which will be used to make the product. The number of decimal places may also be determined by the tolerance of the dimension. For example, a company policy may state that dimensions stated to two decimal places indicate a standard tolerance of ± 0.01 inch, while dimensions stated to three decimal places indicate a tolerance of ± 0.005 inch.

When rounding to a specified number of decimal places, or to a specific place value, calculate the answer to one more place than is specified. Then, use the following rules to round the value.

1. If the digit to the right of the last digit to be kept is less than 5, do not change the last digit to be kept.
    Example: Round 1.243 to 2 decimal places   **1.24**

2. If the digit to the right of the last digit to be kept is more than 5, increase the last digit to be kept by 1.
    Example: Round 1.248 to 2 decimal places   **1.25**

3. If the digit to the right of the last digit to be kept is 5, and there are non-zero digits to the right of the five, increase the last digit to be kept by 1.
    Examples: Round 1.2452 to 2 decimal places   **1.25**

4. If the digit to the right of the last digit to be kept is 5, and there are no non-zero digits to the right of the five, increase the last digit to be kept *only* if it is odd. If the last digit to be kept is even, leave it unchanged.
    Examples: Round 1.2450 to 2 decimal places   **1.24**
    Round 1.2550 to 2 decimal places   **1.26**

ROUNDING USING SIGNIFICANT DIGITS (ACCURACY)

In engineering calculations, the answer to a calculation is often rounded according to mathematical rules for *significant digits*. The use of significant digits is based on the idea that if two numbers used in a calculation are known only to the expressed level of accuracy, the answer can be no more accurate than the least accurate of the original values. The use of significant digit rules are appropriate when the original numbers are approximate values or when the accuracy level of the measurement is indicated in the value. For example, a measurement obtained using a decimal scale with 50 divisions per inch is accurate only to the nearest 0.02" and should be expressed only to that level of accuracy.

The number of significant digits in a number is the number of digits whose accuracy is known or measurable. To determine the significant digits in a number, start at the left with the first non-zero digit and end with the last digit after the decimal point. For example, 20.1, 0.0237 and 1.55 all have 3 significant digits. When rounding to a certain number of significant digits, use one additional digit to round as described above.

A number like 2500 can be ambiguous, since it is not known if the two zeroes are significant or serving as "placeholders" for a rounded value. If it is necessary to show the significant digits, use scientific notation and express the value as 2.5×10^3, 2.50×10^3, or 2.500×10^3 as appropriate.

Example: Round 1.23 to 2 significant digits.
> 1.2 (since 3 is less than five, leave the 2)

Example: Round 1.788 to 3 significant digits.
> 1.79 (change the 8 to a 9 since the next digit is 8)

The use of significant digits to determine how to express a calculated value is guided by two rules:

1. When decimal numbers are *multiplied* or *divided,* their product or quotient should be rounded off to have the same number of significant digits as the factor with the smallest number of significant digits.

2. When decimal numbers are *added* or *subtracted,* their sum or difference should be rounded to the same decimal place as the number with the smallest number of decimal places.

Exceptions to the Rules!

When working with decimal fractions, it is important to use "common sense" and remember the practical situation. For example, if the diameter of a circle is being divided by 2 to find the radius, the use of the significant digit rules would not be appropriate. The 2 is not an approximate value and should not be considered as a "less accurate" value than the diameter.

Example: Find the radius of a circle of 0.56 inch diameter.

$$0.56 \div 2 = 0.28 \text{ inch}$$

Example: Find the radius of a circle of .875 inch diameter.

$$0.875 \div 2 = 0.4375 \text{ or } .438$$

(in practice, a radius dimension is often rounded to the same number of decimal places as the stated diameter)

SCIENTIFIC NOTATION

Many problems in technology and engineering require the use of very large or small numbers. A number is in scientific notation when it is expressed as the product of a number between 1 and 10 ($1 \le$ number < 10) multiplied by a power of 10. The term "power of 10" means that a ten is used as a factor or multiplier the number of times specified by the exponent (the superscript number to the right of 10). For example, $10^2 = 10 \times 10 = 100$. A negative exponent is used to indicate a reciprocal. For example, $10^{-2} = \frac{1}{10^2} = \frac{1}{100}$. (Powers will be discussed further in Unit 23.)

Multiplication by a power of 10 represents a change in the position of the decimal point. For scientific notation, a positive power of 10 indicates that the decimal point in the number is to be moved to the right when converting to standard number form. A negative power of 10 indicates that the decimal point is to be moved to the left.

Examples: $3.25 \times 10^4 = 32,500$

$$3.25 \times 10^{-2} = .0325$$

CALCULATOR USE

Calculators are limited in the number of digits that can be displayed. For example, many scientific calculators display 8 to 10 digits, although internally the calculator may store more digits. One technique to help avoid rounding errors is to enter the entire problem as a series of keystrokes, rather than performing partial calculations and recording the results, which are then re-entered in other calculations. This technique is helpful because the calculator will use the extra digits which are stored in memory in the calculations, rather than just the digits which are displayed.

Note: The value π is often rounded to 3.14 or 3.1416 for calculations. Many calculators also have a key marked π, which is still a rounded value, but is stored internally with more digits than either of those values. If your calculator has a key for π, you may want to use it to increase the accuracy of calculations. In most problems in this text, a value of π will be stated. Your teacher may want you to use the key for π, instead of using 3.14 or 3.1416, and the slight variation in answer will not be significant for most problems in this text.

Scientific calculators typically can be set to display a certain number of decimal places. In that case, most calculators actually round the data to the specified number of decimal places. Some calculators truncate or "drop" extra digits, rather than round. You may want to read the manual for your calculator to see which process it uses. If a calculator is set to display a fixed number of digits, a number can typically be rounded to that number of decimal places by entering the number and pressing the [=] key.

Scientific calculators often have a key marked [Exp] or [EE] which is used to display values with a specified number of decimal places. Some calculators allow you to enter the number of decimal places after the key is pressed, often called fixed point format. Another common feature is engineering notation, in which values are displayed with a power of ten which is a multiple of 3 (10^3, 10^6, 10^{-9}, etc.)

On some calculators, the calculator must be set for either fixed, scientific or engineering format before values are entered. On others, pressing the [Eng] key will cause a number to be displayed with a multiple of 3 power of ten, even though the calculator may be set for a fixed number of decimal places. Be sure that you have read the manual for your calculator and understand how to enter and display values in a specified format.

PRACTICAL PROBLEMS

For problems 1 and 2, round the number to two significant digits.

1. 7.385   _____

2. 5.325   _____

For problems 3 and 4, round the number to three significant digits.

3. 21.368   _____

4. 235.32   _____

For problems 5 and 6, round each number to three decimal places.

5. 548.3128   _____

6. 0.11288   _____

For problems 7 and 8, round each number to two decimal places.

7. 27.375   _____

8. 10.0625   _____

For problems 9 and 10, round each number to the nearest hundredth.

9. 12.4882   _____

10. 32.2144   _____

11. A manufacturing department has the following policy for specifying
     manufacturing tolerances on working drawings:
          2 decimal places in dimension: ± 0.01
          3 decimal places in dimension: ± 0.005
     Using the stated policy, write each dimension in the form 16.70 ± 0.01,
     using the appropriate tolerance.

     a. 32.15   _____

     b. 32.100   _____

     c. 25.325   _____

     d. 27.30   _____

For problems 12–15, each problem was entered on an algebraic logic calculator, and the result
displayed by the calculator is given. Round the answer using significant digit rules.

12. 32.36 × 2.8 = 90.608   _____

13. 27.360 + 8.32 + 12.348 + 2.1 = 50.128   _____

14. 27.36 x 2.34 + 3.225 = 67.2474   _____

15. 5.50 / 4.0 = 1.375   _____

16. Write 0.0575 in scientific notation.   _____

17. Write 3.85×10^4 in standard notation.   _____

18. Write 23,840 in scientific notation.   _____

19. Write 5.625×10^{-3} in standard notation.   _____

20. An angstrom (Å) is a unit of measurement for the lattice parameter for metals, which is a characteristic of the crystalline structure. In scientific notation, one angstrom = 1×10^{-8} cm. Write 1×10^{-8} in standard notation. For each atom listed, write its size in scientific notation and in standard notation.

Atom	Angstroms	Scientific Notation	Standard Notation
aluminum	4.04958		
BCC iron	2.866		
manganese	8.931		

21. One kilowatt-hour (kWh) is equivalent to 3,600,000 joules. Express the number of joules using scientific notation.

_____

22. The length of a microwave is 0.001 meter. Express the length in meters using scientific notation.

_____

23. The cross-sectional area of a ¼" diameter steel rod is calculated by multiplying the radius (.125 inch) by the radius (.125 inch) by 3.14 (π). The calculator displays 0.0490625. Express the area using significant digit rules and include the units of square inches in your answer.

_____

24. Resistivity is a characteristic of a material that describes its resistance to electrical flow. The resistivity of copper at 20° C is 1.724×10^{-6} ohm-cm. Express the resistivity in standard notation.

_____

!25. A store has a sign by a product which says 4/$1.00 or .25¢ each. What is wrong with the sign?

_____

!26. The voltage across the power supply in a series circuit is equal to the sum of the voltage drops across the individual components. If the individual voltage drops are 8.35 volts, 27.3 volts, 3.18 volts, and 3.875 volts, how would the calculated sum of 42.705 V be rounded using significant digit rounding? Explain why.

_____

Unit 12 Addition of Decimal Fractions

BASIC PRINCIPLES OF ADDITION OF DECIMAL FRACTIONS

Adding decimal fractions does not require the use of a common denominator. The process of adding decimal fractions is similar to the addition of whole numbers. The numbers to be added are written in a column, with the *decimal points aligned.* This allows tenths to be added to tenths, hundredths to hundredths, etc. It may be helpful to use zeroes as place holders for numbers which have fewer decimal places than other numbers to be added, as shown in the second illustration.

Example: 24.013 + 3.785 + 4.2 + 3.65

$$
\begin{array}{r}
11\ 1 \\
24.013 \\
3.785 \\
4.2 \\
\underline{3.65\ } \\
35.648
\end{array}
\qquad
\begin{array}{r}
11\ 1 \\
24.013 \\
3.785 \\
4.200 \\
\underline{3.650} \\
35.648
\end{array}
$$

Example: $23.15 + 65¢ + $2.85

$$
\begin{array}{r}
1\ 1 \\
\$\quad 23.15 \\
.65 \\
\underline{2.85} \\
\$\quad 26.65
\end{array}
$$

ADDING NUMBERS EXPRESSED IN SCIENTIFIC NOTATION

When adding numbers expressed in scientific notation, the powers of ten must be the same. Adjust the decimal point and power of ten for each number if needed. Add the numerical values and keep the common power of ten. Then, adjust the sum as needed to express it in scientific notation.

 ## CALCULATOR USE

A calculator can be used to add decimal fractions in a manner similar to adding whole numbers. The decimal point key [.] is pressed at the appropriate place in each number. When using a calculator for addition, it is a common practice to perform the calculations twice and compare the answers. If the answers do not match, a third calculation should be done.

If needed, review the Calculator Use section in Unit 11 for information on entering scientific notation values.

PRACTICAL PROBLEMS

1. 3.9 + 4.5 + 0.3 _____

2. 37.8 + 143.6 + 28.9 _____

3. 1,294.36 + 84.45 + 8.3 _____

4. 542.33 + 8.47 + 2 _____

5. 55.44 + 3.85 + 3.874 + 22.125 + 8.3 _____

6. 5.55 + 22.2 + 3134.8 + 188.56 _____

7. 0.0005 + 1.25 + .075 _____

8. 24.385 + 12.445 + 6.85 _____

9. 1,024.850 + 820.50 + 277.405 _____

10. .0026 + .034 + 0.0025 + 1.3875 _____

11. A feeler gauge is typically marked with a whole number value, which indicates the thickness of the gauge in thousandths of an inch. Several gauges may be "stacked" to measure gaps between parts. What is the total distance (gap) between two engine parts if it is measured with a combination of three feeler gauges marked 10, 2, and 1? (First, write each gauge dimension as a decimal fraction of an inch.) _____

12. A go/no-go gauge is being built to check the size of a drilled hole on a helicopter rotor. The small part of the gauge (minimum dimension) should fit into the cavity, but the large part (maximum dimension) should not. If the part dimension is stated as 1.188 inches ± .015 inch, what is the size of the large end of the gauge? _____

13. The costs of a plumbing repair job included $5.89 for a faucet cartridge kit, $ 0.95 for teflon tape, and $22.50 for labor. What is the total cost of the repair job? _____

14. Find the total length (L) and height (H) of the block illustrated.

L _____

H _____

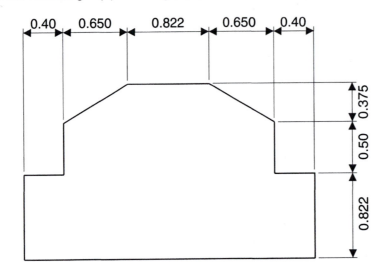

15. A photography student needed two rolls of 36-exposure film and one roll of 24 exposures. If the 36-exposure rolls cost $4.39 each and the 24-exposure roll is $3.47, what is the total cost of the film?

_____

16. For the support block illustrated below, find the missing dimensions.

A _____

B _____

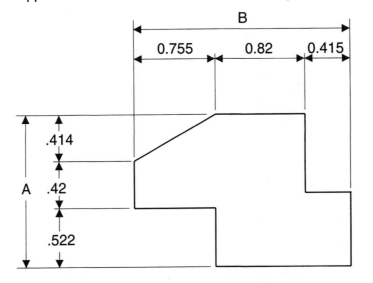

17. A trucking company incurred average costs of $0.26 per mile for the driver, $0.19 per mile for fuel, $0.06 per mile for maintenance, and $0.03 per mile for insurance. What is the total cost per mile to run the truck?

_____

18. A packaging operation for an automobile light requires shrink-wrap film which costs $0.04, a cardboard sheet which costs $0.06, and a label which costs $0.03. What is the cost of each package?

_____

19. A circuit has 3 components connected in parallel. The total current is the sum of the current through the individual branches. If the components are carrying currents of 1.283 amps, 2.015 amps, and 0.788 amp, what is the total amperage?

_____

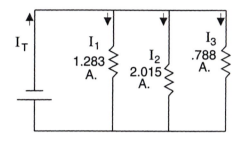

20. Find the total cost of materials if the following are needed for a repair job:

Sheetrock	$45.45
Joint cement	11.25
Paint	17.96
Joint tape	3.49

_____

21. In a parallel circuit, the total current equals the sum of the current in the individual branches. Find the total current (I_T) if the individual currents are $I_1 = 2.455$ amps, $I_2 = 0.84$ amp, and $I_3 = 1.25$ amps.

_____

22. Find the missing dimensions A, B, and C for the mounting plate.

A _____

B _____

C _____

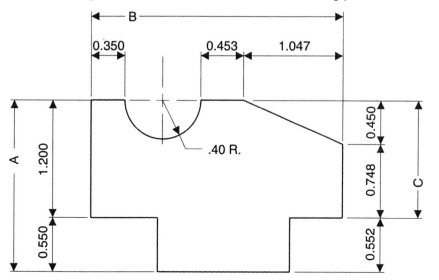

23. A hydroelectric power plant has two units, one which produces 2,400 megawatts of electrical power, and one which produces 3,750 megawatts. The term "mega" means 10^6. Write the output of the power plant in scientific notation and in standard notation.   _____

!24. A technology teacher ordered equipment for a class. Using the order form shown below, find:

  a. the total cost of the items ordered,   _____

  b. total weight of the items,   _____

  c. applicable shipping charges,   _____

  d. and total cost including shipping.   _____

QTY.	NO.	DESCRIPTION	WEIGHT	UNIT COST	COST
1	1124	UNINT. POWER SUPPLY	22 LB.	175.50	
1	3658	DOT MATRIX PRINTER	42 LB.	248.95	
1	4229	DRAFTING TABLE	88 LB.	315.00	
1	8011	AIRBRUSH KIT	8 LB.	277.15	
1	1644	MULTIMETER	4 LB.	65.80	
				SUBTOTAL	

TOTAL SHIPPING WEIGHT		SH.CHG	

SHIPPING CHARGES		TOTAL	
0-10 LB.	$ 5.50		
10.1-50 LB.	$ 9.50		
50.1-100 LB.	$ 14.00		
101-250 LB.	$ 22.50		
OVER 250 LB.	$ 35.00		

!25. Resistors are marked with color bands to indicate their resistance value and tolerance. Reading the value of a resistor requires skills similar to those used in scientific notation. Use the chart in the Appendix to determine the nominal value and tolerance for each resistor described in the chart below.

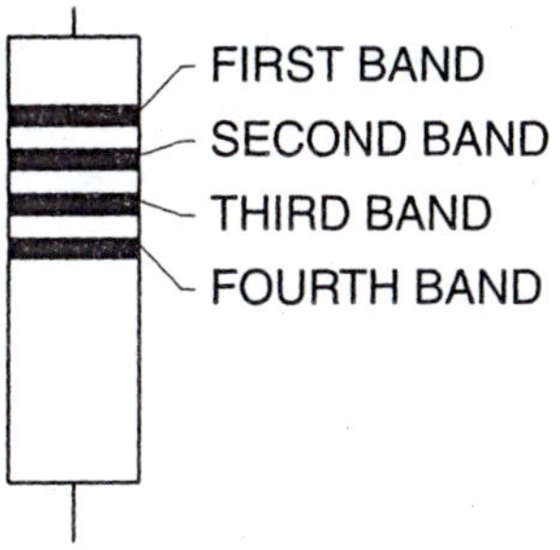

COLOR BAND 1	COLOR BAND 2	COLOR BAND 3	COLOR BAND 4	VALUE	TOL.
RED	BLUE	BROWN	GOLD		
VIOLET	GREEN	BLACK	SILVER		
ORANGE	GREEN	RED	GOLD		

FIRST BAND
SECOND BAND
THIRD BAND
FOURTH BAND

Unit 13 Subtraction of Decimal Fractions

BASIC PRINCIPLES OF SUBTRACTION OF DECIMAL FRACTIONS

The process of subtracting fractions is similar to that of whole number subtraction. The number to be subtracted should be written below the other number, with their decimal points aligned. It may be helpful to add zeroes as place holders at the end of one of the numbers so that both numbers have the same number of decimal places. Subtract the numbers, using regrouping as needed, and place the decimal in the answer directly below the decimal in the numbers.

Example: Subtract 16.23 from 42.385.

$$
\begin{array}{r}
3 \\
42.385 \\
-\ 16.230 \\
\hline
26.155
\end{array}
$$

Example: Subtract 27.852 from 37.56

$$
\begin{array}{r}
26\ \ 5 \\
37.560 \\
-\ 27.852 \\
\hline
9.708
\end{array}
$$

Example: The costs of materials for a job were $27.80 for wire, and $22.57 for electrical plugs and switches. The labor charges were $82.00. If the contractor had estimated that the job would cost $165.00, was the estimate too high or too low? By how much?

$$
\begin{array}{r}
\$\ \ \ 17.80 \\
22.57 \\
82.00 \\
\hline
\$\ \ 122.37
\end{array}
\qquad
\begin{array}{r}
\$\ \ \ 165.00 \\
-\ 122.57 \\
\hline
\$\ \ \ \ \ 42.63
\end{array}
$$

The estimate was $42.63 too high.

SUBTRACTING NUMBERS EXPRESSED IN SCIENTIFIC NOTATION

When subtracting numbers expressed in scientific or engineering notation, make sure that the numbers all have the *same power of 10*. If not, adjust the decimal points and powers of 10 as needed. When the powers of 10 match, align the decimal numbers and subtract. Be sure to include the power of 10 in the calculated answer. In order to write the answer in correct scientific notation, it may be necessary to adjust the decimal point and power of 10 after the calculations are done.

CALCULATOR USE

A calculator can be used to subtract decimal fractions in a manner similar to subtracting whole numbers. The decimal point key [.] is pressed at the appropriate place in each number. To review calculator use for scientific notation, refer to Unit 11.

PRACTICAL PROBLEMS

1. $8.3 - 2.7 - 0.5$ _____

2. $27.85 - 16.25 - 4.375$ _____

3. $155.5 - 3.875 - 45$ _____

4. $122.625 - (37.45 + 44.875)$ _____

5. $(3.8732 \times 10^2) - (8.25 \times 10^{-3})$ _____

6. $8.5 - (2.375 - .9375)$ _____

7. $2.625 - (.0005 + 1.875)$ _____

8. $(1.5785 \times 10^3) - (4.268 \times 10^2) - (2.8 \times 10^{-2})$ _____

9. $37.5 - [22.1 - (.875 + .05)]$ _____

10. $\{55.35 - [79.875 - (22.3 + 44.06)]\}$ _____

11. Gage blocks are being combined to achieve a length of 1.385 inches. If the blocks already connected are 0.700 inch, 0.500 inch, and 0.100 inch, what length needs to be added to achieve the desired length? _____

12. If the clearance between mating engine parts is stated as 0.008" and the inside diameter of one part is 1.325", what is the correct outside diameter for the mating part? _____

13. A construction job required sheetrock costing $23.55, lumber costing $18.30, finishing materials costing $18.95, and $136.00 of labor.

    a. What is the total cost of the job? _____

    b. What is the profit on the job if the customer paid $265.00? _____

14. Tolerance is the total amount of variation allowed in the size of a part. Tolerance is equal to the difference between the maximum and minimum dimensions.

   a. If a dimension is stated as 3.245" ± .004", what is the tolerance? _____

   b. What are the maximum and minimum dimensions?   Minimum: _____

   Maximum: _____

15. Find the missing dimensions on the drawing illustrated below.   A _____

B _____

C _____

D _____

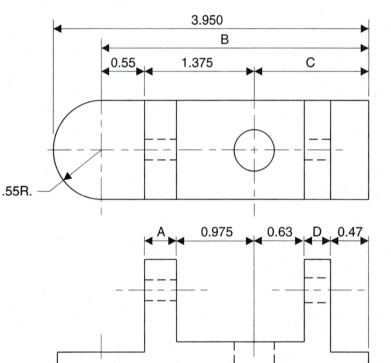

16. a. What is the distance between the centers of holes A and B in the block? _____

    b. What is the distance between the centers of holes B and C? _____

    c. What is the distance between the centers of holes A and C? _____

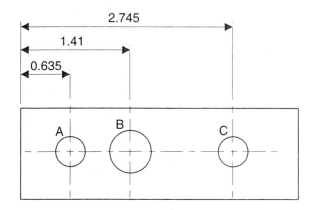

17. A piece of steel 0.880" thick is to be milled to a thickness of 0.825". If the finish cut will remove 0.020", how much material should be removed by the rough cut(s)? _____

18. In a parallel circuit, total current is the sum of the individual (branch) currents. If a circuit with a 15-amp breaker has devices connected in parallel which are carrying currents of 2.53 amps, 5.38 amps, and 4.88 amps, how much more current can the circuit carry before the maximum is reached? _____

19. Dimensions for manufactured parts are often stated as a desired value (the nominal dimension). A tolerance is written after the nominal dimension, as in 0.378 ± .005. Actual part dimensions must fall between the minimum and maximum dimensions, which are calculated by subtracting and adding the specified values to the nominal dimension. Using the stated company tolerance policy based on number of decimal places in the nominal dimension, find the minimum and maximum dimensions for each dimension listed below.

    Dimension tolerance policy:
    Dimensions with 1 decimal place: ± 0.03
    Dimensions with 2 decimal places: ± 0.01
    Dimensions with 3 decimal places: ± 0.004

Nominal Dimension	Minimum Size	Maximum Size
10.3"		
2.375"		
1.88"		
4.625"		
0.45"		
6.425"		
2.95"		

20. An F drill has a diameter of 0.2570", while a $^{17}\!/_{64}$ drill has a diameter of 0.2656". What is the difference in the diameters?

_____

21. Find dimensions A, B, C, and D on the drawing below.

A _____

B _____

C _____

D _____

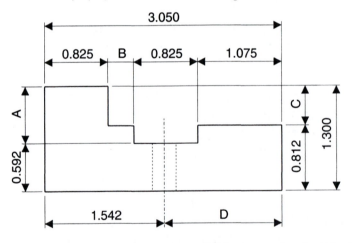

22. The total current in the circuit shown divides at point A. What is the current at point C?

_____

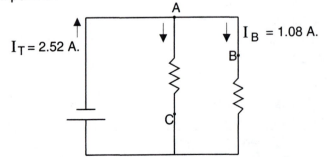

23. The rough opening for a window is required to be 33.625 inches. If the actual window size is 32.8375 inches, what is the difference in the dimensions? _____

24. For the circuit shown, the sum of the voltage drops across the individual components is equal to the voltage across the power source. Find the voltage across resistor 3. _____

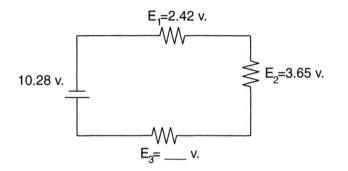

$E_1 = 2.42$ v.

10.28 v.

$E_2 = 3.65$ v.

$E_3 =$ ___ v.

!25. The process plans for a drilling operation call for a size N drill to be used. The tolerance shown on the drawing is ± 0.008.

   a. If the shop has only a fractional drill set, locate the nearest fractional drill in the section of a drill table given and determine whether it would produce a hole within the tolerance specified. _____

   b. Are there any other fractional drills which would be acceptable? _____

Number & Letter Drills	Fractional Drills	Decimal Equivalent (inch)
	9/32	.2812
L		.2900
M		.2950
	19/64	.2969
N		.3020
	5/16	.3125
O		.3160
P		.3230
	21/64	.3281

Unit 14 Multiplication of Decimal Fractions

BASIC PRINCIPLES OF MULTIPLICATION OF DECIMAL FRACTIONS

The process of multiplying decimal fractions is similar to the multiplication of whole numbers. However, the multiplication of decimal fractions requires that a decimal point be located correctly in the product. The number of decimal places in the product is the total number of decimal places in both numbers being multiplied. For example, if one number has one decimal place and the other has two, the product will have three decimal places. After the calculation is done, the answer may be rounded to a specific number of decimal places or by using significant digit procedures as needed.

Example: Multiply 44.375 × 2.5 and round to 2 decimal places.

$$
\begin{array}{r}
1\,1 \\
2\,1\,3\,2 \\
4\,4.3\,7\,5 \quad \text{(3 decimal places)}\\
\times \quad\quad 2.5 \quad \text{(1 decimal place)}\\
\hline
2\,2\,1\,8\,7\,5 \\
8\,8\,7\,5\,0 \quad\;\; \\
\hline
1\,1\,0.9\,3\,7\,5 \quad \text{(4 decimal places)}\\
\text{or } 1\,1\,0.9\,4 \quad \text{(2 decimal places)}
\end{array}
$$

MULTIPLYING NUMBERS EXPRESSED IN SCIENTIFIC NOTATION

When values to be multiplied are expressed in scientific notation, multiply the numeric values together. The power of 10 will be the sum of the powers of 10 in the factors. Note that the *signs* of the exponents must be considered. Addition of signed numbers is an important algebra concept. Basically, when adding signed numbers, if the signs of the numbers are the same, add the numbers and keep the same sign. If the signs are different, find the difference of the numbers and keep the sign of the larger number.

Example: $(3.48 \times 10^2) \times (2.51 \times 10^4)$

$= (3.48 \times 2.51) \times (10^{2+4})$

$= 8.7348 \times 10^6$

Example: $(3.48 \times 10^5) \times (2.51 \times 10^{-2})$

$= 8.7348 \times 10^{(5-2)} = 8.7348 \times 10^3$

CALCULATOR USE

A calculator can be used to multiply decimal fractions. Enter each number, using the decimal point [.] key at the appropriate location in each number. The multiplication key is normally marked with an **x**. To multiply two numbers on an algebraic logic calculator, enter the first number, press the multiplication key, enter the second number, and press the equal [=] key. Remember the Order of Operations when entering problems with multiple steps or operations.

PRACTICAL PROBLEMS

Do not round answers on problems 1–10.

1. 3.5×1.8 _____

2. 25.78×9.30 _____

3. $100.37 \times .0045$ _____

4. $1,254.8 \times 35.375$ _____

5. $(.25)(37.25 - 4.75)$ _____

6. $(3.87 \times 10^3) \times (4.5 \times 10^2)$ _____

7. $(1.06 \times 10^{-3}) \times (87.5 - 42.3)$ _____

8. $(.58)(2.5 \times 2.5 \times 3.14)$ _____

9. $[2.25 - (3.5 \times 10^{-2})(5.4 \times 10)]$ _____

10. $(0.5)(50.0 \times 27.3 \times 27.3)$ _____

11. According to Ohm's Law, E (voltage) = I (amperage) x R (resistance). If a circuit with a total resistance of 55.5 ohms is carrying a current of 0.40 amps, what is the voltage?   _____

12. Two students are employed while attending a technical school. David earns $4.75 per hour at a fast food restaurant. Sam earns $6.90 per hour as a junior drafter.

    a. If each student worked 42 hours during a 2-week pay period, how much did each student earn?   David/Sam _____

    b. What is the difference in their earnings per hour? _____

    c. What is the difference in their earnings for the pay period? _____

    d. Assuming that each student worked 20 hours per week for 36 weeks, find each student's earnings   David/Sam _____

    e. Find the difference in the earnings. _____

13. A hydroelectric power plant has three units, each producing 1,250 megawatts.

    a. With all three units running, what is the total output of the power plant? _____

    b. Write the answer in both standard notation and scientific notation. _____

14. An industrial engineer has developed a way to reduce 0.25 hour from the time required to assemble a product.

    a. How much time would be saved on a production run of 640 products? _____

    b. If the hourly cost for that assembly operation is $20.85, what is the total savings from reducing the assembly time on that run? _____

15. A machinist earning $14.85 per hour worked 36 hours.

    a. Find her total earnings. _____

    b. If the deduction for income tax is found by multiplying 0.12 by total earnings, what is the income tax deduction? _____

    c. If the deduction for Social Security and Medicare is determined by multiplying .0835 by the gross pay, find the amount of that deduction. _____

    d. What is the net (take-home) pay after deducting income tax and Social Security and Medicare taxes? _____

16. An industrial supply company has a price schedule based on the number of items purchased at one time. The schedule is illustrated in the chart below. A manufacturing company uses 1,300 total parts per year and wants to compare the cost of ordering all of them at once versus ordering in 2 smaller orders.

AAA Industrial Supply Co.
Price List

Part No.	Description	1-500	501-900	900 +
100-86	⅝" valve repair kit	$1.82	$1.74	$1.58
101-35	½" I.D. bearing	$0.62	$0.56	$0.43
104-52	⅝" O-ring set	$0.14	$0.13	$0.11

   a. If the company orders 500 of part 101-35 in January and another 800 in July, what is the total cost of the parts? _____

   b. If the company orders 1,300 in January and stores 800 of them at a cost of $0.07 each until they are needed in July, what is the total cost? _____

   c. Which purchasing option is less expensive and what are the savings using that option? _____

17. A drafter earning $9.25 per hour works for four 8-hour days on a drawing.

   a. How much does the drawing cost the company in labor costs? _____

   b. If the company also provides a benefit package estimated to be 0.16 of the hourly wage, what is the drafter's total equivalent hourly wage? _____

   c. How much is invested in the drawing, including the benefit package? _____

18. According to Watt's Law, P (power) = E (voltage) × I (amperage). If a circuit has a current of 2.5 amps with a voltage of 117.3 volts, find the power (P) in watts. _____

19. A box pattern requires a 1.25 square foot piece of sheetmetal. What is the cost of the material if the sheetmetal costs $0.60 per square foot? _____

20. A manufacturing process produces 0.35 pound of scrap aluminum per part.

    a. If 280 parts are made, what is the weight of scrap produced?   _____

    b. If the scrap can be sold to a recycler for $0.37 per pound, what is the amount received for selling the scrap?   _____

21. A machinist earning $10.65 per hour worked for 36 hours.

    a. What were her gross wages?   _____

    b. If the total tax deduction is 0.18 times the gross wage, how much is deducted for taxes?   _____

22. The energy consumed by a 1,100 watt (1.1 kW) toaster operated for 1 hour would be 1.1kW ×1 hr = 1.1 kWh. If electric energy costs 7.5 cents per kWh, find an hourly cost to operate the toaster.   _____

23. An automobile produces about 0.4 pound of nitrogen oxides for every gallon of gasoline burned.

    a. If an average car uses 750 gallons of gasoline per year, how many pounds of nitrogen oxides are produced by that car in a year?   _____

    b. If a town has 30,000 cars, what is the amount of nitrogen oxide they produce annually?   _____

24. A dining room light has five 60-watt bulbs. If electric energy costs 0.009 cents per watt-hour, find the cost to operate the light for 4 hours.   _____

25. The R-value of a certain insulation material is 3.33/inch.

    a. What is the R-value of 3.5" thick batt insulation?   _____

    b. What is the R-value of 5.5" thick batt insulation, to the nearest whole unit?   _____

26. If a furniture repair project needs 7.5 square feet of veneer at $3.60 per square foot, what is the cost of the veneer?   _____

Unit 15 Division of Decimal Fractions

BASIC PRINCIPLES OF DIVISION OF DECIMAL FRACTIONS

Division of decimal fractions is done in basically the same way as division of whole numbers, except that the decimal must be handled and placed correctly in the quotient. Place the divisor to the left of the division bracket, with the dividend under the bracket, as in whole number division. Change the divisor into a whole number by moving the decimal point to the right end of the number. Then move the decimal in the dividend the same number of decimal places to the right, adding zeroes if needed. The decimal point in the quotient is then aligned directly above the "new" decimal point in the dividend. The numbers are divided in the same way as whole numbers.

Example: Divide 36.54 by 5.8

$$
\begin{array}{r}
6.3 \\
58\,\overline{)\,365.4} \\
\underline{34\ 8} \\
17\ 4 \\
\underline{17\ 4} \\
0
\end{array}
$$

Example: Divide 127.05 by 8.25

$$
\begin{array}{r}
15.4 \\
8.25\,\overline{)\,127.05\ 0} \\
\underline{82\ 5\ \ \ \ } \\
44\ 45 \\
\underline{41\ 25\ \ } \\
3\ 30\ 0 \\
\underline{3\ 30\ 0} \\
0
\end{array}
$$

Example: A car used 15.2 gallons of gas to travel 407.4 miles. What was its fuel consumption in miles per gallon to the nearest tenth?

407.4 mi ÷ 15.2 gal = 26.8 miles per gallon

Unlike whole number division, the quotient in decimal fraction division does not include a remainder. If the division does not end with a remainder of zero, add zeroes to the dividend as needed and continue dividing until the appropriate number of decimal places is achieved in the quotient. The quotient may need to be carried to one more place than the desired precision and then rounded.

DIVIDING NUMBERS EXPRESSED IN SCIENTIFIC NOTATION

When values to be divided are expressed in scientific notation, divide the decimal values. To find the correct power of 10, subtract the power of 10 in the divisor from the power of 10 in the dividend. Remember that the signs of the exponents must be considered.

Example: $(15.24 \times 10^4) \div (3 \times 10^3)$

$$= (15.24 \div 3) \times (10^{4-3})$$

$$= 5.08 \times 10^1 = 5.08 \times 10$$

(Note that an exponent of 1 does not have to be written.)

Example: $(12.8 \times 10^{4)} \div (2 \times 10^{-2})$

$$= (12.8 \div 2) \times (10^{4-[-2]})$$

$$= 6.4 \times 10^{4+2} = 6.4 \times 10^6$$

 CALCULATOR USE

To divide decimal fractions on an algebraic logic calculator, enter the dividend, including the decimal point. Then, press the division key [÷], and then enter the divisor. Press the equal [=] key to display the quotient. On calculators, the answer to a division problem (the quotient) is normally displayed to several decimal places. The number may terminate or repeat. The answer may need to be rounded to a specified accuracy or precision.

PRACTICAL PROBLEMS

1. $65.85 \div 3$ _____

2. $5.7 \div 3.8$ _____

3. $32.428 \div 2.2$ _____

4. $553.5 \div 15.375$ _____

5. $(4.58 \times 10^3) \div (2.5 \times 10)$ _____

6. $(55.30 - 12.8) \div (3.625 - 1.125)$ _____

7. $65 \div 60 \div 60 \times 5,280$ _____

8. $(0.5)(524.8) \div (32.5 - 12)$ _____

9. $(1.375 \times 10^3) \div (2.2 \times 10)$ _____

10. $(1.4875 \times 10^2) \div (7 \times 10^{-2})$ _____

11. A car is driven 385 miles, using 13.2 gallons of gas. Find the fuel efficiency of the car in miles per gallon. (Note: The term "per" indicates division, with the unit "miles per gallon" indicating division of the number of miles by the number of gallons.) _____

12. Ohm's Law states that R (resistance) = E (voltage) ÷ I (amperage). If the voltage in a circuit is 110 volts and the amperage is 5.5 amps, find the resistance in ohms (Ω). _____

13. Watt's Law states that P (power) ÷ E (voltage) = I (amperage). If P = 1,150 watts and the voltage is 110 volts, find the amperage. Round your answer to the nearest 0.1 amp. _____

14. What is the hourly rate of pay if Steven receives $38.70 for 4.5 hours of work? _____

15. The term "equally spaced" occurs often on drawings for manufactured parts. What is the distance between the centers of adjacent holes in the figure below? _____

ADJUSTMENT PLATE FOR SUPPORT FIXTURE

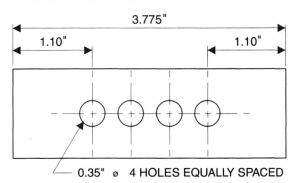

0.35" ø   4 HOLES EQUALLY SPACED

16. Printer ribbon test data shows that a name-brand ribbon lasts an average of 560 pages and costs $10.40. A generic ribbon lasts an average of 420 pages and costs $8.20. Which has the lower average cost per page? (Note: For comparisons of this type, more digits are normally used for comparison than standard significant digit practice would use. Calculate the costs per page to the nearest tenth of a cent or thousandth of a dollar.) _____

17. A roll of insulation has 77.5 square feet of batts and costs $19.37. What is the cost per square foot to the nearest cent?

_____

18. Material in a vocational school is sold to students at cost. If a 50' roll of wire costs $12.60, what is the cost per foot to the nearest tenth of a cent?

_____

19. The return on an investment is calculated by dividing the profit by the amount invested. If a company invests $248,000 in a new product and earns $8,680 in profit, what is the return on investment to the nearest thousandth? (Note: This type of information is often stated as a percentage; similar problems will be studied in Section 5.)

_____

20. A multiview drawing of a block is shown.

   a. If the views are to be positioned on the paper so that the horizontal spaces between borders and views are to be equal, find the spacing distance.

   _____

   b. If the views are to be positioned so that all spaces are equal vertically, what is that spacing distance?

   _____

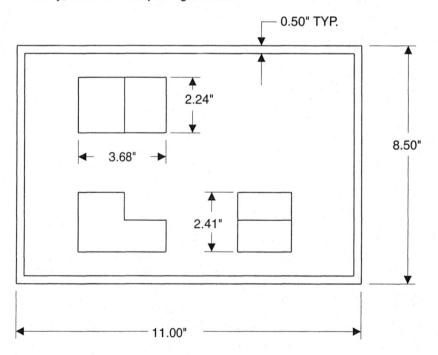

21. The time intervals between sequential parts moving on an automated assembly line were recorded as 3.40 seconds, 1.88 seconds, 2.22 seconds, 2.28 seconds, 3.01 seconds, and 1.97 seconds. What is the average time between parts? _____

22. Watt's Law states that P (power) / E (voltage) = I (amperage). If P = 1,382 watts and the voltage is 216 volts, find the amperage. Round your answer to the nearest 0.1 amp. _____

23. A copier uses a spring to hold a light mechanism in place. The spring exerts a force (F) of 20 pounds when stretched 1.75 inches from its original length (*l*). When the technician replaces the spring, what spring constant (k) should be specified? Use the formula k = F ÷ *l*. _____

24. When a nut is placed on a ⅜-16 bolt, one turn of the nut moves it 0.0625" down the length of the bolt. How many turns are needed to move the nut 0.4375" down the bolt? _____

25. Support posts for a power supply cabinet are cut from 36" long pieces of plastic extrusion.

    a. If each support post is 1.15" long and each cut requires 0.03", how many posts can be cut from each piece? _____

    b. What is the length of the remaining piece? _____

26. A circuit contains 4 identical lamps, connected in series. The total voltage across the power supply (7 volts) is equal to the sum of the voltage drops across the individual lamps.

    a. Write a math expression using division which could be used to find the voltage drop across each lamp. _____

    b. What is that voltage drop? _____

Unit 16 Decimal and Common Fraction Equivalents

BASIC PRINCIPLES OF DECIMAL AND COMMON FRACTION EQUIVALENTS

It is important to be able to convert numbers into different forms for various applications. Although the form looks different, the value of the number should be the same, with the exception of rounding errors which sometimes occur in decimal fractions.

Example: $\frac{14}{1} = 14$ $\frac{3}{3} = 1$ $\frac{16}{2} = 8$ $\frac{5}{10} = \frac{1}{2}$

Changing a Common Fraction to a Decimal Fraction

A common fraction, which is used to express a part of a whole unit, can also be thought of as expressing division. A common fraction can be changed into its decimal equivalent by dividing the numerator by the denominator, as shown in the examples below. (Note that if the denominator of a common fraction has prime factors other than 2 or 5, the decimal form will be a repeating decimal.)

$$\frac{5}{8} = 5 \div 8 = 0.625$$

$$\frac{11}{32} = 11 \div 32 = 0.34375$$

$$\frac{1}{3} = 1 \div 3 = 0.333 \text{ (the 3 repeats)}$$

Converting a Mixed Number to Decimal Form

A mixed number can be converted to its decimal equivalent by converting the fraction part to its decimal form and adding it to the whole number.

Example: $2\frac{5}{8} = 2 + \frac{5}{8} = 2 + 0.625 = 2.625$

Changing a Decimal Fraction to a Common Fraction

A nonrepeating decimal fraction can be written as a common fraction with a denominator that is a power of 10. The power of 10 in the denominator is the same as the number of decimal places in the decimal fraction.

Example: $0.28 = \frac{28}{100}$

Another way to convert a decimal fraction to its common fraction form is to "read" the decimal fraction and then write the fraction which would be read the same way. For example, 0.385 is read as "three hundred eighty-five thousandths" and is written as $385/1000$.

Converting a Decimal to the Nearest "Ruler Fraction"

A decimal can be converted to the nearest "ruler fraction," meaning that it has a denominator of 2, 4, 8, 16, etc. Multiply the decimal form of the fraction by the desired denominator and round the number to the nearest whole number. Use that number as the numerator in a fraction with the selected denominator.

Example: Write 0.660 as a fraction to the nearest eighth.
$0.660 \times 8 = 5.28$ Write the fraction $5/8$.

Example: Write 0.38 as a fraction to the nearest sixteenth.
$.38 \times 16 = 6.08$ Write the fraction $6/16 = 3/8$.

CALCULATOR USE

When using a calculator to convert a fraction to decimal, enter the numerator, press the divison [÷] key, enter the denominator, and then press the equal [=] key.

PRACTICAL PROBLEMS

1. Convert $5/8$ to decimal form.   _____

2. Convert $9/32$ to decimal form.   _____

3. Convert $2^{11}/64$ to decimal form.   _____

4. Convert $12 1/4$ to decimal form.   _____

5. Convert 0.4 to fraction form.   _____

6. Convert 0.375 to fraction form.   _____

7. Convert 2.025 to fraction form.   _____

8. Convert 0.8375 to fraction form.   _____

9. a. Convert 0.33 to the nearest fraction with a denominator of 16.   _____

    b. Then, write the next smaller fraction and the next larger fraction with
        denominators of 16.   _____

    c. Convert each to decimal form and verify that the original decimal is
        between those two values.   _____

10. For each of the numbers in the table below, write the prime factors of the
    number. Then, write the word "terminate" or "repeat" in the next column
    to indicate whether a fraction with that denominator would terminate or
    repeat when converted to decimal form.

Denominator	Prime Factors	Terminate or repeat?
5		
8		
10		
16		
7		
12		
9		
24		
64		
32		

11. Convert 0.842 to the nearest fraction with a denominator of 8.   _____

12. For the drawing below, convert the fraction dimensions to their equivalent decimal form. Express each decimal dimension to the nearest hundredth of an inch.

$3\frac{1}{2} =$ _____

$2\frac{7}{8} =$ _____

$\frac{5}{8} =$ _____

$\frac{5}{16} =$ _____

$\frac{7}{16} =$ _____

$\frac{11}{16} =$ _____

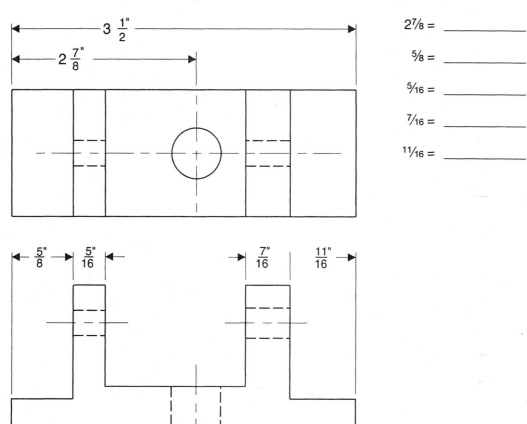

13. A part drawing calls for a 0.400" diameter hole to be drilled through the part. If only fractional size twist drills are available, which of the drills listed in problem 14 is the closest to the desired dimension without being larger than the specified diameter?

_____

14. Some common twist drill sizes are listed below. Complete the chart by filling in the missing values to 4 decimal places.

Drill Size Conversion Chart

Drill	Decimal Size
5/64	
3/16	
9/32	
3/8	
13/32	
7/16	
5/8	
3/4	

15. A set of drawings for a shelf unit was dimensioned in decimal inches. Since the carpenter will use a standard fractional steel tape to construct the bookcase, the dimensions need to be converted to fraction form. Using the drawing, calculate the fractional form of each dimension to the nearest 1/16" increment.

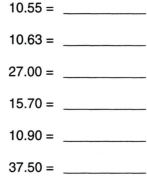

10.55 = _____

10.63 = _____

27.00 = _____

15.70 = _____

10.90 = _____

37.50 = _____

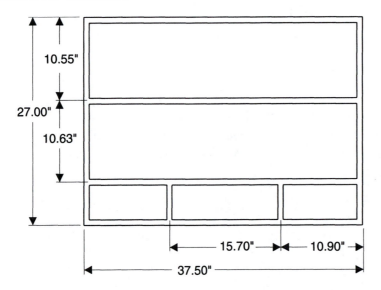

SHELF UNIT FOR ROOM 114 - TECHNOLOGY BUILDING

16. A quality control inspector must determine if a machined dimension is within tolerance. She uses a dial caliper to check the part, so the actual sizes are measured in thousandths of an inch. The part dimensions are stated as ⅞" ± ¹⁄₆₄". Express the minimum and maximum allowable sizes in decimal form to the nearest thousandth of an inch. Then determine which of the parts listed on the checksheet are acceptable (within tolerance).   Acceptable sizes: _____

Quality Control Dept.			
Inspection Report			
Part No. 1035-2		Dim. Size 7/8 + 1/64" - 1/64"	
I.D.	Measured Dim.	Accept/Reject	Date
186	0.883"		
187	0.879"		
188	0.892"		
189	0.863"		
190	0.875"		
191	0.878"		
192	0.857"		
193	0.888"		

17. A ⅞" diameter shaft must fit into a sleeve that is to be 0.005" larger in diameter than the shaft. What is the inside diameter of the sleeve, expressed in decimal form?   _____

18. An aluminum plate was 1.875" thick before being machined. After machining, the thickness was 1.500". Find the amount of reduction and express it in fraction form.   _____

19. A finishing operation is to be done on a 1⅝" diameter part. If the finishing operation adds 0.003" to the outside surface, what is the finished diameter expressed as a decimal?   _____

20. A student needs to make a chart showing decimal and fraction equivalences to memorize common conversions. Fill in the missing conversions on the chart below. For repeating decimals, round to 3 decimal places. (Hint: You should memorize common conversions, also!)

Fraction	Decimal
$\frac{1}{16}$	
$\frac{1}{8}$	
$\frac{1}{6}$	
$\frac{1}{5}$	
$\frac{1}{4}$	
$\frac{1}{3}$	
$\frac{1}{2}$	

21. One ounce is $\frac{1}{16}$ of a pound.

a. What fraction of a pound is 5 ounces?   _____

b. What is the decimal form of the fraction?   _____

22. Fill in the blanks in the following statement with the decimal form of the fraction to the nearest thousandth.

The standard allowance for machine reaming is $\frac{1}{64}$" (_____) for holes under $\frac{1}{2}$" (_____) diameter and $\frac{1}{32}$" (_____) for holes which are $\frac{1}{2}$" and larger.

23. A student is making a chart showing decimal and fraction equivalences to memorize. Fill in the missing conversions on the chart below. For repeating decimals, round to 3 decimal places.

Fraction	Decimal
$\frac{2}{3}$	
$\frac{3}{4}$	
$\frac{5}{6}$	
$\frac{3}{8}$	
$\frac{5}{8}$	
$\frac{7}{8}$	
$\frac{7}{16}$	

!24. Convert the following fractions to decimal form.

Fraction	Decimal
¼	
⅞	
⁵⁄₁₆	
¹⁄₃₂	
½	
¹¹⁄₆₄	

Determine whether a "rule" or pattern exists which relates the denominator to the number of decimal places in the decimal form. (Try a few more fractions if needed.)

a. If all decimal values are to be expressed to thousandths of an inch, which ruler denominators will have to be rounded?   _____

b. Which will have to have zeroes added as placeholders?   _____

!25. Write the fractions ¹⁰⁄₆₄ and ⁸⁄₃₂ as decimals. Do they fit the rule or principle you determined in problem 23? Write a brief statement explaining why or why not. What qualification(s) must be added to your rule to make sure that it always predicts the number of decimal places?

Unit 17 Review and Combined Operations
with Decimal Fractions

Practical Problems

1.   35.478 + 8.54 − 17.2 + 3.005 _____

2.   6.5 ÷ 4 + 3 × 2.75 _____

3.   4.74 + 3.8 + 5.2525 + 0.01 _____

4.   27.5342 × 3.14 _____

5.   Express 5.2854×10^{-2} in standard notation. _____

6.   Express 3,254.5 in scientific notation. _____

7. $(2.545 \times 10^2) + (3.385 \times 10^3)$ _____

8. $[21.3 - (3.5 \times 10^{-2})(7.2 \times 10)]$ _____

9. $(4.27 \times 10^2) \times (2.453 \times 10)$ _____

10.   (0.5)(25)(4.8)(4.8) _____

11.   What is the gap between engine parts if the distance is measured using
     a combination of feeler gauges which are 0.010", 0.002", and 0.001"
     thick? _____

12.   Total amperage in a parallel circuit is the sum of the amperages of the
     different components. A circuit currently has a clock using 0.027 amp, a
     coffeemaker with 4.86 amps, and an electric mixer with 1.87 amps. If a
     toaster using 7.3 amps is turned on, will a 15-amp circuit breaker be
     sufficient to carry the current load? _____

13.   A motor shaft is ¾" in diameter. The diameter of the bearing into which
     the shaft fits is 0.782" with a tolerance of ± .003".

     a.   What is the minimum clearance? _____

     b.   What is the maximum clearance? _____

14. Fill in the missing dimensions on the drawing shown.

A _____

B _____

C _____

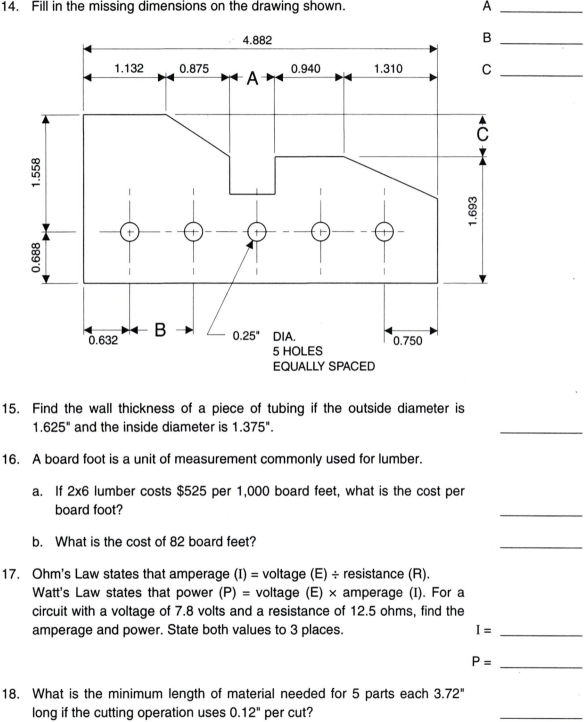

0.25" DIA.
5 HOLES
EQUALLY SPACED

15. Find the wall thickness of a piece of tubing if the outside diameter is 1.625" and the inside diameter is 1.375".

_____

16. A board foot is a unit of measurement commonly used for lumber.

a. If 2x6 lumber costs $525 per 1,000 board feet, what is the cost per board foot?

_____

b. What is the cost of 82 board feet?

_____

17. Ohm's Law states that amperage (I) = voltage (E) ÷ resistance (R). Watt's Law states that power (P) = voltage (E) × amperage (I). For a circuit with a voltage of 7.8 volts and a resistance of 12.5 ohms, find the amperage and power. State both values to 3 places.

I = _____

P = _____

18. What is the minimum length of material needed for 5 parts each 3.72" long if the cutting operation uses 0.12" per cut?

_____

19. A carpenter estimates that a cabinet will require 56 board feet of lumber at $1.86 per board foot, three hours of labor at $10.50 per hour, 1 quart of stain at $6.70, and 1 qt. of polyurethane finish at $6.90. What is the estimated total cost? _____

20. The R-value of fiberglass insulation is reported to be 3.33 per inch of thickness. How many inches are required to provide an R-value of at least 16? (Round your answer to the nearest inch.) _____

21. An employee worked 36 hours at $8.65 per hour. His gross wages are reduced by taxes (0.16 × wages) and retirement fees (0.08 × wages). Calculate the following:

    a. His deductions for taxes. _____

    b. His deduction for retirement. _____

    c. His net pay. _____

22. A student purchases 20 bolts at 6 cents each, 20 washers at 3 cents each, 20 nuts at 4 cents each, and 6 feet of chain at 37 cents per foot.

    a. What is the cost of the materials? _____

    b. If the sales tax is calculated by multiplying the cost by 0.0825, what is the amount of tax? _____

    c. What is the total cost including tax? _____

23. The charge to have a copier repaired includes a trip charge and labor costs. The trip charge is $25.00 for travel up to 50 miles plus 15 cents per mile over 50 miles, and labor is $38.00 per hour. Find the total charge if the trip is 68 miles and the repairman works for 1.5 hours. _____

24. Watt's Law states that I (amperage) = P (power) ÷ E (voltage). Ohm's Law states that R (resistance) = E (voltage) ÷ I (amperage). If P = 72 Watts and E = 120 volts, find I (amps) and R (ohms). I _____

    R _____

25. If a production run has a fixed cost of $850 for the run and variable costs of $8.75 per part, what is the total cost for a run of 2,150 parts? _____

26. A nightlight has a 7-watt bulb.

 a. If electricity costs $0.10 per kilowatt-hour, find the cost of operating the light continuously for 24 hours. _____

 b. Calculate a weekly cost to run the light 24 hours a day. _____

 c. Find the cost for operating the light for 30 days. (Remember that 1k=1,000. To convert watts to kilowatts, divide by 1,000.) _____

27. An endmill which was originally 1.250" in diameter was resharpened and now measures 1.238" in diameter.

 a. Find the reduction in diameter expressed in decimal inches. _____

 b. If the endmill must be moved over (offset) ½ of the reduction amount to maintain the proper cutting position, what is the required offset? _____

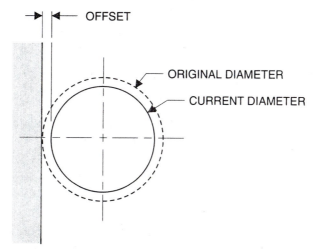

28. The dimension for a part is stated as 1.325 (+.010/−.006). What are the minimum and maximum dimensions?

Maximum _____

Minimum _____

29. If a house requires 33 squares of shingles, what is the cost difference between 3-tab shingles at $19.65 per square and dimensional shingles at $36.50 per square?   _____

30. A class determined that a project requires 18" of 2x4 or 12" of 2x6 for each item produced. The class plans to produce 40 items. The class needs to decide whether to purchase 2x4's or 2x6's. If the cost of a 12' 2x4 is $4.89 and a 12' 2x6 costs $5.95,

   a. how many 2x4's would be needed? Will there be any extra?   _____

   b. how many 2x6's would be needed? Will there be any extra?   _____

   c. what is the cost of that many 2x4's?   _____

   d. what is the cost of that many 2x6's?   _____

   e. which is the better buy?   _____

   f. what is the difference in cost between the two options?   _____

Ratio and Proportion

A ratio is a way of comparing two or more values. Ratios are used extensively in Industrial Technology applications, including scaled measurements, roof pitch, mechanical advantage, and efficiency of systems, and in materials technology (alloys and chemical mixtures or solutions, for example.)

A proportion is a way of stating that two ratios are equivalent. Proportions are useful in calculating unknown values that can be expressed in a ratio.

In this unit, ratios will be written to express relationships in values, and proportions will be written and solved for missing values.

Unit 18 Ratios

BASIC PRINCIPLES OF RATIO

A ratio is a way of comparing two numbers. It is normally written either as a fraction *a/b* or in the form *a:b*. Note that the order of the terms in a ratio is important. When a written description of a ratio is expressed in mathematical symbol form, the value before the word "to" is written as the numerator or to the left of the colon.

A ratio should be written in its lowest terms, like a common fraction. Regardless of whether it is written as a fraction or with a colon, divide out any common factors of the two values to reduce it to lowest terms.

Example: The ratio of wheels to steering wheels on an automobile is 4:1 or 4/1.

You may also be familiar with the ratio of the circumference (C) of a circle to its diameter (d), which is a constant (π), often approximated as 3.14. $C/d = \pi$

Several ratios can be written for the same data. For example, if there are 12 men and 15 women in a class, ratios which could be written include men to women (12:15), women to men (15:12), men to total students (12:27) or women to total students (15:27). The ratio which is written depends on the intended use of the ratio.

It is also possible to compare more than two values using a ratio. For example, concrete is prepared by mixing aggregate (gravel), sand, and cement with water. One ratio which is sometimes used, 5 parts aggregate to 3 parts sand to 1 part cement, can be written as 5:3:1. The term "parts" refers to the chosen measurement unit.

Inverse Ratios

An inverse ratio is the ratio written in reverse order. The inverse ratio of 3:5 is 5:3. When a ratio is written in fraction form, the inverse ratio is its reciprocal. The inverse ratio of 5/8 is 8/5.

Comparing Ratios

To determine whether two ratios are equal, write each in lowest terms and compare. Another way to determine whether they are equal is called *cross-multiplication.* Write each ratio in fraction form. Multiply the numerator of one ratio by the denominator of the other. Repeat using the other numerator and denominator. If the two products are equal, the ratios are equal. This concept will be used in Unit 19 to solve proportions.

Example: Is the ratio 3:5 equal to 15:28?

$$\frac{3}{5} \overset{?}{=} \frac{15}{28} \qquad 3 \times 28 = 84$$
$$5 \times 15 = 75$$

Since the cross-products are not equal, the ratios are not equal.

PRACTICAL PROBLEMS

For problems 1–4, write each ratio in lowest terms.

1. 162 : 27 _____

2. 33 : 132 _____

3. 2.7 : 1.5 _____

4. 5/8 : 1/8 _____

For problems 5–8, write the inverse ratio.

5. 150 : 29 _____

6. 15 : 22 _____

7. 4.3 : 1 _____

8. 3/15 : 7/16 _____

9. Are the ratios 21:28 and 3:4 equal? _____

10. Are the ratios 2/15 and 32/45 equal? _____

11. A school has 384 students and 24 teachers.

    a. Write a ratio for the number of students to the number of teachers. _____

    b. Write a ratio for the number of students to the total number of students and teachers. _____

12. A gear ratio is a comparison based on the number of teeth on each gear. It is stated as

$$\frac{\text{teeth on driven gear}}{\text{teeth on driver gear}}$$

If the driver gear has 12 teeth and the driven gear has 48 teeth, what is the gear ratio?   _____

13. Mechanical advantage for a simple machine is defined as the ratio of output force to input force. If 20 pounds is applied to lift a load of 60 pounds, what is the mechanical advantage of the machine used?   _____

14. On a belt driven machine, the spindle is turning 350 revolutions per minute (rpm) while the motor is turning 1,750 rpm. Write a ratio of motor speed to spindle speed.   _____

15. A computer lab at a high school has 15 computers with a 486 chip, and 11 computers with a 586 chip. Output devices include 2 laser printers, 2 drum plotters, and 3 dot matrix printers. During the school year, 234 students used the lab.

   a. Write a ratio comparing the number of 486 computers to the number of 586 computers.   _____

   b. Write a ratio comparing the number of students to the number of computers.   _____

   c. Write a ratio comparing the number of computers to the number of output devices.   _____

16. Bronze is an alloy of copper and tin. If a quantity of the alloy contains 7 oz of tin and 70 oz of copper, what is the ratio of copper to tin?   _____

17. Brass is a copper alloy with 10–45% zinc by weight.

   a. For 150 lb of brass containing 45 lb of zinc, write a ratio expressing the weight of zinc to total weight of brass.   _____

   b. Then, write a ratio expressing the weight of copper to total weight of brass.   _____

   c. Write a ratio expressing the weight of Zn to the weight of Cu.   _____

18. A consumer research survey on a new car model reported that for every 100 cars sold, 72 problems were reported. What is the ratio of problems to cars sold?

_____

19. A sample of 200 parts is inspected and 12 are found to be defective.

a. What is the ratio of defective to total inspected parts?

_____

b. What is the ratio of defective to acceptable parts?

_____

20. In a model bridge building competition, one measure of quality is the structural efficiency, which is a ratio of the failure load of the bridge to the weight of the bridge. If one of the model bridges carried a load of 36 pounds and weighed 2 pounds, what was its structural efficiency?

_____

21. A transformer has 65 secondary turns and 520 primary turns. What is the ratio of primary turns to secondary turns?

_____

22. The U.S. 5-cent coin is made of 1 part nickel (Ni) and 3 parts copper (Cu). Write a ratio expressing the amount of nickel to total metal in the coin.

_____

23. The mechanical advantage of a simple machine is defined as the ratio of output force to input force. If a force of 42 N (newtons) is applied to a pulley system to lift a load of 231 N, what is the mechanical advantage?

_____

24. Slope of a line is a ratio of the vertical rise to the horizontal run of the line. If the rise is 7.5 meters and the run is 18 meters, what is the slope?

_____

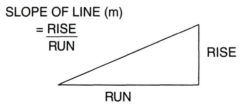

SLOPE OF LINE (m)
$$= \frac{RISE}{RUN}$$

RISE

RUN

25. The pitch of a roof is a way of stating its slope. In standard form, a roof pitch is stated in fraction form, with a denominator of 12 feet. For a roof which rises 8¾ feet over a run of 21 feet, find the slope and convert it to standard pitch form (with a denominator of 12).

_____

26. Simple machines have a "trade-off" between force and distance over which the force is applied. A force of 35 N was applied to a lever to lift a load of 280 N. The distance over which the input force was applied was 32 cm, while the load was lifted 4 cm.

a.   Write the ratio of output force to input force.   _____

b.   Write a ratio comparing output distance to input distance.   _____

c.   What is the relationship between the two ratios?   _____

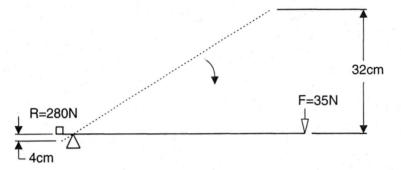

For problem 27 and in other problems in this text, a special notation is used to distinguish between different values such as voltage across different components. Subscript notation, such as R_T, is common in Industrial Technology problems. The subscript T is normally used to indicate a total value, while a number subscript is used to describe a characteristic of a single component.

27. For the circuit shown, R_1 is a 20 Ω (ohm) resistor and R_2 is a 40 Ω resistor. The voltage drop across R_1 is 40 volts and the voltage drop across R_2 is 80 volts.

a.   Write a ratio comparing the voltage drop across R_1 to the voltage drop across R_2.   _____

b.   Then, write a ratio comparing the resistance in ohms of R_1 to the resistance of R_2.   _____

c.   What is the relationship between the two ratios?   _____

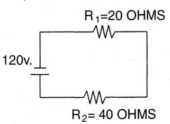

28. A measurement scale commonly used for residential buildings is ¼" = 1'0" (or ¼" = 12"), meaning that ¼" on the drawing represents 1' or 12" of the building.

   a. Write a ratio of drawing length to building length. (Be sure to use the same units such as feet or inches.)

         _____

   b. Then, convert the fraction to an equivalent fraction with a numerator of 1 to find the actual size reduction (for length) of the drawing.

         _____

29. Concrete for a foundation was mixed using 5 parts of cement to 12 parts of sand to 25 parts of aggregate. Is that ratio the same as a cement-to-sand-to-aggregate ratio of 1:3:5? (Hint: First, compare the cement-to-sand ratios and then the sand-to-aggregate ratios.)

         _____

30. For the circuit shown, find the total resistance (R_T) of the circuit using the relationship $\frac{1}{R_T} = \frac{1}{R_1} + \frac{1}{R_2}$. The voltage across the power supply (E) is 8 volts. Find the total current (I_T) at point A using the relationship $I_T = \frac{E}{R_T}$.

At point B, the current splits, with part of the current traveling each path. If the voltage across each resistor (E_1 and E_2) is also 8 volts, find the current through each resistor using the relationship $I_n = \frac{E_n}{R_n}$, where n is the number of the component. Write a ratio comparing the resistance of R_1 to the resistance of R_2. Then, write a ratio comparing the current (I_1) through R_1 to the current (I_2) through R_2. What is the relationship between the two ratios?

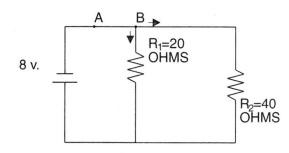

R_T _____

I_T _____

I_1 _____

I_2 _____

$R_1{:}R_2$ _____

$I_1{:}I_2$ _____

_____

Unit 19 Proportion

BASIC PRINCIPLES OF PROPORTION

A proportion is a statement that two ratios are equivalent. Both ratios in the proportion must be written in the same order, such as miles to gallons. A proportion can be written in two ways, depending on which way the ratios are expressed. Both ways are shown below.

$$a : b :: c : d \qquad \frac{a}{b} = \frac{c}{d}$$

Example: If 4 feet of aluminum extrusion are needed to produce 10 parts, 12 feet will be needed to produce 30 parts.

$$\frac{4}{10} = \frac{12}{30} \qquad 4 : 10 :: 12 : 30$$

Direct Proportion

Values are in direct proportion to one another when, as one increases, the other increases.

Inverse Proportion

Values are in inverse proportion to one another when, as one increases, the other decreases.

SOLVING A PROPORTION FOR A MISSING VALUE

If a proportion is written in fraction form, it can be solved using *cross-multiplication.* Rather than using cross-multiplication to determine whether two ratios are equal, using cross-multiplication to solve a proportion involves stating that the ratios (one including an unknown value or variable) are equal. The diagonal values (numerator of one ratio and denominator of the other) are multiplied and placed on one side of an equation. The equation is then solved for the missing value.

Example: If 45 feet of cable costs $9.90, what would 85 feet of the cable cost?

$$\frac{45'}{\$9.90} = \frac{85'}{x}$$
$$45'\, x = \$9.90\,(85')$$
$$x = \$18.70$$

If a proportion is written using colons and the symbol (::), the two values on the far left and right ends of the proportion are known as the *extremes*. The two values next to the :: sign are known as the *means*. To solve the proportion, set the product of the means equal to the product of the extremes. Solve the equation for the missing value.

Example: If 45 feet of cable costs $9.90, what would 85 feet of the cable cost?

$$45 : 9.90 \; :: \; 85 : x$$
$$45\,x = 85\,(\$9.90)$$
$$x = \frac{(\$9.90)\,(85)}{45}$$
$$x = \$18.70$$

 CALCULATOR USE

When solving a proportion for a missing value, enter the cross-product which contains two numbers first. Obtain the product, then press the division key [÷] and enter the number from the cross-product which contains a variable.

PRACTICAL PROBLEMS

For problems 1–10, solve each proportion for the unknown value.

1. $\dfrac{2}{7} = \dfrac{x}{56}$ _____

2. $3 : 14 :: 12 : x$ _____

3. $\dfrac{1.5}{2.4} = \dfrac{x}{8.64}$ _____

4. $\dfrac{12}{66} = \dfrac{18}{x}$ _____

5. $\dfrac{1}{4} : \dfrac{3}{4} :: \dfrac{2}{3} : x$ _____

6. $\dfrac{840}{6} = \dfrac{18,480}{x}$ _____

7. $6 : 14 :: 15 : x$ _____

8. $51.3 : 3.42 :: x : 1$ _____

9. $27{,}432 : 108 :: x : 65$ _____

10. $22.5 : 3{,}780 :: 3 : x$ _____

11. A sample of 134 parts was drawn randomly from a production lot of 2,680 items. There were 4 defective parts in the sample. Assuming that the sample is representative of the lot, how many defective items would be expected in the entire lot?   _____

12. If a pollution count is stated as 312 particles per cubic meter, and a room contains 504 cubic meters of air, what is the predicted number of particles in the room air?   _____

13. Bronze is an alloy of copper and tin, typically with a 10:1 copper to tin ratio.

   a. How much tin should be combined with 32 lb of copper to make a 10:1 bronze?   _____

   b. What is the weight of the bronze produced?   _____

14. A common application of proportions is for similar triangles. Using the statement that "Corresponding parts of similar triangles are proportional," determine the missing values in the diagram below. Round to the nearest tenth.

x = _____

y = _____

a = _____

b = _____

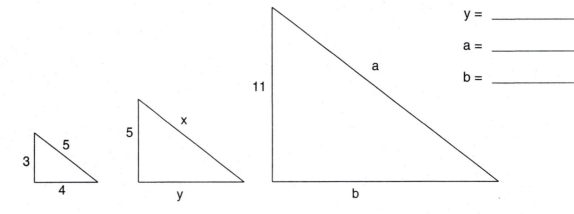

SIMILAR TRIANGLES

15. A developing tank requires 21 oz of solution to develop film. If the mixing ratio of chemical to water is 1:2, how many ounces of chemical and of water should be mixed? (Hint: First, write a ratio of chemical to mixed solution.)

chemical: _____

water: _____

16. On a trial production run, 22 parts were produced in 4.5 hours. At the same rate, how long, to the nearest hour, should it take to produce 1,600 parts?

_____

17. A hole is to be drilled vertically in the block, with the center of the hole located 1" from the top, measured down the inclined surface. Using the principle of similar triangles, what is the horizontal displacement (x) needed for the drill position?

_____

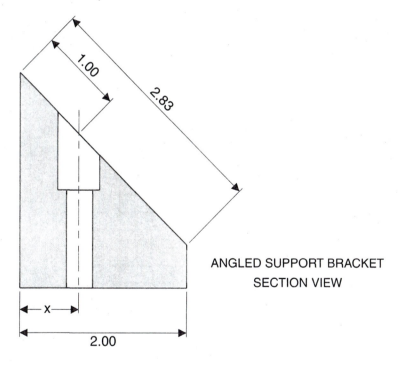

ANGLED SUPPORT BRACKET
SECTION VIEW

18. A 3:4:5 triangle is often used to "square" a building, using the concept of similar triangles. If the carpenter measures 15 feet down one edge of the building (for the 3 in the ratio), how far should he measure down the other side and across the diagonal?   _____

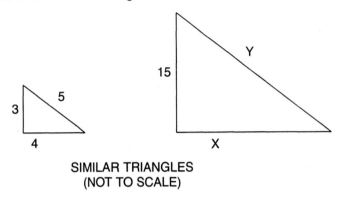

SIMILAR TRIANGLES
(NOT TO SCALE)

19. Slope is an important concept in math. It is defined as a ratio of vertical rise to horizontal run for a line segment.

$$m = \frac{rise}{run}$$

Find the slope of lines A, B, C, and D shown below.

A _____

B _____

C _____

D _____

20. The specifications for a drainage pipe state that the pipe must drop 1" for every 4' of length of the pipe. How many inches of drop are needed for a pipe 116 feet long? _____

21. Resistance in a wire is proportional to the length of the wire. The resistance of 100' of a wire is 0.0035 ohms. What is the resistance for 580' of the same wire? _____

22. A pump is able to move 188 liters of cutting fluid from a storage tank to a machine tank in 8 minutes. How long should it take to move 2,250 liters from the storage tank to another tank at the same rate? Round to the nearest minute. _____

23. A building 82' long was drawn at a scale of ¼" = 1'0".

    a. Express the scale as a ratio of drawing unit to actual unit. _____

    b. Determine the length of the drawing. _____

    c. If the width of the building was 47', what width would it be drawn? _____

24. If 85' of reflective tape costs $25.35, what would be the cost for 550' of tape at the same rate? _____

25. Roof pitch is a way of stating the slope of a roof. If a roof rises 12¼' over a horizontal span of 21', what is its pitch? _____

26. The relationship between input and output forces and the distances over which they must be applied in simple machines can be stated as $\frac{E}{R} = \frac{RD}{ED}$. If a force (E) of 35 N is applied over a distance (ED) of 32 cm to lift a load (R) of 280 N, how far (RD) will the load be lifted? _____

27. In oxyacetylene welding, acetylene and oxygen gases are used to produce a high-temperature flame. The ratio of oxygen pressure to acetylene pressure is typically 5:2.

    a. Although it is not recommended, if the acetylene were used at its maximum stable pressure of 15 pounds per square inch (psi), what oxygen pressure would be used according to the ratio? _____

    b. If the acetylene pressure were set at its recommended maximum pressure of 10 psi, what oxygen pressure would be needed? _____

28. For a roof with a pitch of 5.5/12, what would be the height for a horizontal run of 18.6 feet? _____

29. If 35 parts come off of an assembly line conveyor in 12.6 seconds, how long should it take to get 1,500 parts? _____

30. When one type of concrete block is installed, the distance from the center of a mortar joint to the next joint is 16", so that the distance covered by 3 blocks is 4 feet.

    a. Write a ratio comparing a distance of 4 feet to 3 blocks. _____

    b. If the size of a room is to be estimated by counting the blocks, what size (in feet) is a room which "measures" 28½ blocks long and 24 blocks wide? _____

Unit 20 Combined Problems in Ratio and Proportion

1. Write the ratio 15:48 in lowest terms. _____

2. Write the ratio 42/360 in lowest terms. _____

3. Write the inverse ratio of 450:7. _____

4. Are the ratios 42:360 and 40:288 equal? _____

5. Are the ratios 1:3:7 and 18:54:126 equal? _____

6. Solve $\frac{2}{3} : 1 :: \frac{3}{4} : x$ for the unknown value. _____

7. Solve $\frac{1,680}{5} = \frac{x}{1.5}$ for the unknown value. _____

8. Solve $\frac{3}{4.8} = \frac{x}{17.28}$ for the unknown value. _____

9. Solve $0.6 : 1 :: 14.4 : x$ for the unknown value. _____

10. In the proportion $\frac{E}{R} = \frac{RD}{ED}$, are R and RD directly or inversely proportional? Explain how you determined your answer. _____

11. Mechanical advantage is defined as a ratio of output force to input force. What is the mechanical advantage of a pulley system if a 45-pound force is needed to lift a 126-pound load? _____

12. A sample of 16 parts is drawn from a production lot of 200 parts and inspected. If 3 of the parts are defective, what is the estimated number of defective parts in the lot, assuming that the sample is representative of the lot? _____

13. The instructions for mixing cutting fluid are to add ½ liter of concentrate to 38 liters of water. If a quantity of 22 liters of cutting fluid is to be prepared, how many liters of concentrate and water should be used? Express your answer to the nearest tenth. _____

14. On a trial production run, 4 parts were produced with individual production times of 6.3, 5.8, 5.9, and 6.2 hours.

    a. What is the total trial production time?   _____

    b. If the trial run is used to prepare a bid for 26 additional parts, what is the estimated total production time for those 26 parts?   _____

15. As described in problem 17 of unit 18, brass is an alloy of copper and zinc.

    a. To make brass which is $35/100$ zinc, how much copper and zinc are needed to make 45 lb of brass?   _____

    b. How much of each would be needed for 1,200 lb of the same type of brass?   _____

16. The ratio of output voltage to input voltage for a transformer is the same as the ratio of secondary coil turns to primary coil turns, ignoring any transformer losses. If a transformer has 576 turns on the secondary winding and 120 windings on the primary, and the input voltage is 9 volts, what is the output voltage?   _____

17. The instructions for mixing a chemical solution for photographic developing state that 2 ounces of chemical concentrate are to be mixed with 32 ounces of water.

    a. What is the ratio of concentrate to water?   _____

    b. What is the ratio of chemical concentrate to total solution?   _____

    b. If 170 ounces of solution are needed, how many ounces of water and chemical concentrate should be mixed?   _____

18. A sample of 240 parts was drawn randomly from a production lot of 3,000 parts. There were 4 defective parts in the sample. Assuming that the sample is representative of the lot, how many defective items would be expected in the entire lot?   _____

19. The motor of a drill press is running at 1,750 rpm and the spindle is turning at 1,050 rpm.

    a. What is the ratio of motor speed to spindle speed? _____

    b. If the belt/pulley system which determines the ratio is unchanged, and the motor speed is increased to 2,400 rpm, at what speed would the spindle turn? _____

20. A residential structure was drawn at a scale of ¼" = 1'0" (or ¼" = 12"). If the length of the building is 58 feet and the width is 42 feet, find the length and width of the drawing.

    L _____

    W _____

21. If a 15 N force is applied to a simple machine to lift a 120 N load, what is the mechanical advantage of the machine? _____

!22. A formula used to convert from Celsius to Fahrenheit degrees is stated as $F = \dfrac{9}{5}C + 32$. Find out the number of degrees between the freezing and boiling points for water on the Fahrenheit scale. Find out the number of degrees between the freezing and boiling points for water on the Celsius scale. In which scale is a degree a larger unit? Write a ratio comparing the number of degrees between boiling and freezing on the Fahrenheit scale to the number of degrees between those points on a Celsius scale, and reduce the ratio to lowest terms. Write a brief statement explaining the purpose of the "9⁄5" factor in the conversion formula.

    _____

    _____

    _____

    _____

    _____

Percents, Averages, and Estimates

Percentages are used often in technical and engineering problems. They are involved in such topics as interest paid or earned, discounts allowed, cost overruns or savings, and estimates prepared for jobs. The relationship between fractions, decimal fractions, and percent will be emphasized.

Averages are also important in technical problems, along with the ability to estimate answers to problems. The term "estimate" can be used two ways with reference to problems in engineering and technology:

1. a process used to obtain an approximate cost for a job (such as in construction), or

2. a way of using simple, approximate values to obtain an answer, rather than working with more precise but more complicated calculations.

Examples of both uses of the term "estimate" will be included in this section.

Unit 21 Percent and Percentages

BASIC PRINCIPLES OF PERCENT AND PERCENTAGES

The term *percent* means *per hundred* or *divided by 100*. (It may help to remember that a *cent* is ¹⁄₁₀₀ of a dollar). Using that definition, 35% can be written as the common fraction ³⁵⁄₁₀₀ or as the decimal fraction 0.35.

When a percent sign is written after a number, it represents a division by 100. In order to keep the value of the number the same, the decimal must be moved to the right two places (multiplying by 100). The number looks different, but its value is unchanged.

After completing this unit, you should be able to express numbers in decimal, fraction, or percent form as needed for different applications. Note that 100%=1 and any percent greater than 100% will be a number greater than one when it is converted to decimal or fraction form.

Example: 120% = 1.20 = 1 ⅕   200% = 2.00

Changing a decimal fraction to a percent

To change a decimal fraction to a percent, move the decimal point two places to the right and add a percent sign.

Example: Express 0.85 and 0.125 as percents.

$$0.85 = 85\%$$
$$0.125 = 12.5\%$$

Changing a percent to a decimal fraction

To change a percent to a decimal fraction, move the decimal point two places to the left and omit the percent sign.

Example: Express 67% and 150% as decimal fractions.

$$67\% = 0.67$$
$$150\% = 1.50$$

Changing a common fraction to a percent

Changing a common fraction to a percent involves two conversions. The common fraction should first be changed into a decimal fraction by dividing the numerator by the denominator. Then change the decimal fraction into a percent by moving the decimal two places to the right and adding a percent sign.

Example: Express ⅝ as a percent.

⅝ = 0.625 = 62.5%

Changing a percent to a common fraction

Changing a percent to a common fraction also involves two conversions. First, change the percent to a decimal fraction by moving the decimal two places to the left and omitting the percent sign. Then change the decimal fraction into a common fraction (described in Unit 16) and reduce to lowest terms.

Example: Change 28.4% to its fraction form.

$28.4\% = 0.284 = {}^{284}/_{1000} = {}^{71}/_{250}$

Expressing one number as a percent of another number

To express one number as a percentage of another number, divide the first number by the second number. This procedure is often used in calculating grades as a percentage. For example, if a student gets 21 of 25 questions correct, the percentage correct is determined by dividing 21 by 25 (which is equal to 0.84) and converting it to a percent form (84%).

${}^{21}/_{25} = 0.84 = 84\%$

Calculating a specified percentage of a number

To find a specified percentage of a number, write the percentage in decimal form and multiply by the number.

Example: Find 24% of 86.

24% = 0.24   0.24 × 86 = 20.64

USING RATIOS AND PROPORTIONS TO WRITE PERCENT PROBLEMS

Percent problems can be "set up" by considering them as proportion problems, involving four quantities: the base (total amount), the part (part of the base), the percentage rate, and the value 100. The relationship is expressed as $\dfrac{\%}{100} = \dfrac{part}{base}$. Substitute the known values into the proportion and solve for the unknown.

Example: Find 12% of 850.

$$\frac{12}{100} = \frac{x}{850}$$
$$(12)(850) = (100)(x)$$
$$10{,}200 \div 100 = x = 102$$

Example: 90 is 15% of what number?

$$\frac{15}{100} = \frac{90}{x}$$
$$15x = (90)(100) = 9{,}000$$
$$x = 9{,}000 \div 15 = 600$$

PRACTICAL PROBLEMS

1. Convert 0.88 to percent. _____

2. Convert 2.35 to percent. _____

3. Convert 74% to decimal form. _____

4. Convert 25.8% to decimal form. _____

5. Convert ⅞ to percent. _____

6. Convert 1¼ to percent. _____

7. Convert 220% to a mixed number. _____

8. Convert 85% to a fraction. _____

9. Convert 62.5% to a fraction. _____

10. Convert ⅞ to a decimal and to percent form. _____

11. 45 is what percent of 250? _____

12. What number is 27 percent of 2,400? _____

13. 105 is what percent of 3,500? _____

14. 238 is 35 percent of what number? _____

15. 533,750 is what percent of 854,000? _____

16. A company normally includes a 2.5% cost adjustment factor in its bids to cover price increases in materials during construction. If the material is estimated to cost $253,000, what is the amount of the cost adjustment factor to be added? _____

17. An automotive technician received a raise from $8.50 per hour to $9.25 per hour after completing an advanced training course. What percent of his original salary was the raise? _____

18. Four CNC lathes were used to produce table legs in a furniture factory. Before a programming error on one lathe was corrected, 64 legs were produced with a slightly different contour. What percent of the total run of 400 legs was different? _____

19. The efficiency of a system is an important characteristic of the system. It is defined as the ratio of work out of the system to work input to the system. If the work input to a pulley system was 2,400 N-m and the output was 1,425 N-m, find the efficiency and express it as a percentage. _____

20. How many pounds of sulfur are contained in 2,250 pounds of coal having 4% sulfur content? _____

21. A company implemented an energy savings plan, with the result that the average electric bill for one year was $357 per month, compared with $432 per month the previous year. Express the average cost with the savings plan as a percentage of the previous average. _____

22. A woodworking project is estimated to have a 15% waste factor due to knots and imperfections in the wood. If the plans require 42 board feet, how much should be ordered for the project, including the estimated waste? _____

23. Manufacturing cost for a motor was estimated to be $22.80. The actual manufacturing cost was $20.90. The actual cost was what percentage of the estimated cost? Round your answer to the nearest percent.   _____

24. Of a shipment of 4,400 parts, 186 were found to be defective. What is the percentage defective? (Round to the nearest tenth of a percent.)   _____

25. The maintenance costs of a robot are estimated to be 5% of its cost the first year and 8% of its cost the second year. If the robot cost $128,000.00, find the estimated maintenance costs for the first and second years.

    First: _____

    Second: _____

26. In the AISI and SAE four-digit steel designation system, the last 2 digits give the carbon content in *hundredths of a percent*. For example, 1040 is $^{40}/_{100}$% or 0.4% carbon. For 4130 steel, find the percent of carbon, and then convert it to decimal form.   _____

27. An industrial supply company charges 5% of the order total as a shipping/handling charge. If the order total is $876.35, what is the shipping charge?   _____

28. High grade aluminum ore is 45% alumina. If 32,000 pounds of ore are mined, how many pounds of alumina can be obtained?   _____

29. Nickel silver is 55% Cu, 15% Ni, and the rest is Zn.

    a. What percent of the nickel silver is Zn?   _____

    b. In a 180-pound quantity of nickel silver, find the weight of Cu, Ni, and Zn in the alloy.

    Cu _____

    Ni _____

    Zn _____

30. Heat loss through cracks and poor-fitting doors and windows can account for 15% of a homeowner's energy bill. A homeowner's bills for the year are shown in the chart below.

   a. What is the total energy bill for the year? _____

   b. If the wasted energy due to cracks, etc., is 15%, how much is spent for the wasted energy? _____

   c. What percent of the total annual energy costs are incurred during the winter season (December, January, and February)? _____

Month	Amount
January	$ 135.20
February	$ 140.82
March	$ 95.50
April	$ 88.60
May	$ 86.20
June	$ 92.57
July	$ 102.38
August	$ 128.96
September	$ 108.10
October	$ 79.25
November	$ 84.48
December	$ 105.88

31. If a product was estimated to cost $34.20 to manufacture, but an 8% cost overrun was incurred, what was the final manufactured cost of the product? _____

32. A company is required to pay an 8.25% sales tax on supplies which are used rather than being resold. A sales receipt for supplies purchased in January is shown below. Find the missing values to fill the blanks.

```
┌──────────────────────────────────────────────────────┐
│              INDEPENDENT HARDWARE                      │
│                 ANYTOWN, U.S.A                         │
│                                                        │
│        Copier paper                     $  24.50       │
│            2 @ 12.25                                    │
│        Pencils                          $   5.34       │
│            6 @ .89                                      │
│        Black fine-tip markers           $   2.76       │
│            4 @ .69                                      │
│        Computer paper                   $  46.05       │
│            3 @ 15.35                                    │
│        Printer ribbons                  $  48.40       │
│            5 @ 9.68                                     │
│                                                        │
│            Subtotal taxable          _____         │
│                                                        │
│        Tax @ 8.25%                   _____         │
│                                                        │
│        Total due                     _____         │
│                                                        │
└──────────────────────────────────────────────────────┘
```

33. A steel alloy (chromium-molybdenum 4130) has 0.70% Cr and 0.15% Mo additives. Convert each of the percentages to decimal form.   0.70% _____

    0.15% _____

34. An aluminum casting has 0.9% shrinkage by volume during the solidification phase and 0.6% shrinkage by volume during cooling in the solid state.

    a. If the casting has an initial volume of 85 cubic inches, find the amount of shrinkage during solidification and the volume after solidification.   _____

    b. Then find the amount of shrinkage during solid-state cooling in cubic inches. (For shrinkage during solid-state cooling, use the volume after solidification shrinkage as the initial volume.) Record volumes and shrinkage amounts to 3 decimal places.   _____

35. An industrial engineer conducted a time/motion study and recommended changes to reduce the amount of time needed for an assembly operation. The original time was 18.2 seconds, while the modified time was 15.7 seconds. Express the reduction as a percent of the original time. Round to the nearest tenth of a percent.

    _____

36. An energy conversion process actually involves three conversion steps. If the individual efficiency of step 1 is 80%, of step 2 is 75%, and of step 3 is 70%, what is the overall efficiency? (Note: To determine overall efficiency, multiply the individual efficiencies.)

    _____

Unit 22 Simple Interest

BASIC PRINCIPLES OF INTEREST

One application of percentages is in calculating interest, which is the charge for borrowing money or the amount received for having money invested or in savings. There are two types of interest: simple and compound. In this section, problems involving simple interest will be worked. Compound interest will be presented in Unit 39 (Solving Equations). Compound interest is more common today.

Basic Principles of Simple Annual Interest

An *interest rate* is expressed as a percentage, and is usually stated as an *annual rate.* The amount of money borrowed or invested is called the *principal.* The amount of interest (I) charged or earned is found by multiplying the principal (P) by the annual interest rate (i) by the appropriate length of time (t) in decimal form. $I = (P)(i)(t)$

When the loan is repaid, the amount paid (A) is the sum of the principal and interest ($A = P + I$).

Example: A small company borrows $10,000 (P) to purchase computer equipment. The annual interest rate (i) is 14% and the money is borrowed for one year. How much interest is charged for the year? What is the total amount of money repaid?

$P \times i = I$

$10,000 \times 0.14 = $1,400 interest

$P + I = A$

$10,000 + $1,400 = $11,400 total amount paid

Example: A computer repair technician invested $8,050 in a savings account for 2 years. The annual interest rate was 6%. Find the amount of interest earned.

$I = ($8,050)(0.06)(2) = 966

Daily and Monthly Interest Rate Calculations

Today, interest is often calculated using a daily interest rate, as seen on many credit card statements. The daily rate is obtained by dividing the annual interest rate by 365 and carrying the quotient to several decimal places. In other instances, it may be stated as a monthly rate by dividing the annual rate by 12. The daily or monthly rate is then multiplied by the number of days or months and then multiplied by the principal.

Example: A company had a loan with a credit union which charged interest based on daily calculations. If the account balance was $3,450 for a 28-day period and the annual interest was 12%, find the interest charge. (Round the daily interest rate to 6 decimal places.)

0.12 / 365 = .000329 daily interest rate

.000329 × $3,450 × 28 = $31.78 interest

Declining Balance Loans

Many loans today, such as mortgages, credit card loans, or automobile loans have interest charged each month. The new balance each month is calculated by subtracting payments from the previous balance, and adding any new charges and interest. Interest is calculated on a monthly basis.

Example: A drafting company purchased a computer at a cost of $2,750. The annual interest rate was 9%, with $355 monthly payments. Calculate the interest charges and balance for the first 4 months.

DATE	PREVIOUS BALANCE	+ INTEREST (P)(.09)÷12	– PAYMENT	NEW BALANCE
January	$ 2750.00	$ 20.63	$ 355.00	$ 2415.63
February	2415.63	18.12	355.00	2078.75
March	2078.75	15.59	355.00	1739.34
April	1739.34	13.05	355.00	1397.39

PRACTICAL PROBLEMS

For problems 1–4, calculate the simple interest and total payoff for the loan amounts specified.

1. Principal of $1,800, annual interest rate of 8% for 3 years _____

2. Principal of $5,000, annual interest rate of 14% for 1 year _____

3. Principal of $85,000, annual interest rate of 9% for 20 years _____

4. Principal of $11,000, annual interest rate of 8% for 5 years _____

For problems 5 and 6, calculate the daily interest rate to 6 decimal places and the interest which would be charged on the specified loan amount for the time stated.

5. Principal of $3,250, annual interest rate of 15% for 30 days _____

6. Principal of $24,000, annual interest rate of 7.9% for 28 days _____

For problems 7 and 8, calculate the monthly interest rate to 4 decimal places and the interest which would be charged on the specified loan amount for the time stated.

7. Principal of $1,500, annual interest rate of 9% for 6 months _____

8. Principal of $12,000, annual interest rate of 16.5% for 3 months _____

For problems 9 and 10, calculate the total payoff for the loan amounts and rates stated.

9. Principal of $2,680, annual interest rate of 12% for 2 years _____

10. Principal of $4,500, annual interest rate of 18.9% for 1 year _____

11. A company invested $85,000 in an investment which paid 8% annually. If the investment was held for 2 years, how much interest was earned? _____

12. An electrical supplier billed a contractor for $1,680 in parts on January 3. If the interest rate was 18% annually and the contractor paid the bill on March 3, find the interest charge if the interest is calculated monthly. _____

13. A company borrowed $5,200 for 3 months to purchase a new lathe.

   a. If the annual interest rate is 10%, what is the monthly interest rate to 5 decimal places? _____

   b. How much interest would be charged on the loan? _____

14. An electrical supply company shipped an order of wire and breakers to a lumber yard. The bill for $1,650 was due within one month to avoid interest. If the bill was not paid within one month, interest would accrue at the rate of 18% annually from the billing date. The bill was paid after two months. Find the amount of interest charged and the total amount to be paid. _____

15. A carpenter invested $12,000 in a bank, earning 5% simple interest per year. If the money is invested for 6 months, how much interest will be earned?   _____

16. A carpenter borrowed $525 to finance an electric miter saw at 9.5% annual simple interest. How much interest will be charged at the end of one year?   _____

17. A contractor arranged interim financing for a house at 10.8% annually and withdrew funds ("draws") as shown below. Find the amount of interest charged for each time period, using a monthly interest rate and the previous balance. Then find the new balance after each draw, adding the interest charge and the draw to the previous balance. Use September 1 as the payoff date. (Note: On June 1, find the interest on the money borrowed from April 1 to June 1, etc.) Then find the total interest charged and the final payoff amount.   Total interest: _____

Final payoff: _____

DATE	PREVIOUS BALANCE	NUM. OF MONTHS	INTEREST	DRAW	NEW BALANCE
4/1	xxxxx	xxxxx	xxxxx	$18,000	$18,000
6/1	$18,000			$23,500	
8/1				$32,000	
9/1				xxxxx	

18. The balance on a credit card account is calculated by deducting payments from the previous balance and adding interest and other charges. For the statement below, fill in the missing values. (Calculate the interest using the previous balance for each month. Use the new balance for one month as the previous balance for the next month.) The interest rate on the account is 15% annually.

DATE	PREV. BAL.	– PAYMENT	+ INTEREST	NEW BAL.
3/1	1,850.00	200.00		
4/1		100.00		
5/1		150.00		
6/1		50.00		

Unit 23 Discount

BASIC PRINCIPLES OF DISCOUNT

Discounts are another common use of percentages in business. Discounts are sometimes given for payment of a bill within a specified period of time, for purchasing items in larger quantities, for members of a particular trade, or a variety of other reasons. The original price is called the *list price,* while the discounted price is called the *net price.* To determine the amount of a discount, express the discount percentage *as a decimal* and multiply it by the price to which it is to be applied.

Discount (D) = Discount percent (p) × List price (L)

Net price (N) = List price (L) − discount (D)

Single Discount

To calculate a discount, multiply the discount rate by the list price. To find the net price, subtract the amount of the discount from the list price.

Example: A computer repair company offers a 5% discount if an invoice is paid within 30 days. If the invoice is for $276.40, what is the amount of the discount? What is the net price?

$276.40 × 0.05 = $13.82   discount

$276.40 − $13.82 = $262.58   net price

An alternate method for finding the net price is to subtract the discount percent from 100% to obtain the percent of the net price to be paid. The decimal form of that percent is multiplied by the original price.

Example: If a discount of 5% is to be applied to a purchase of $276.40, what is the net price?

100% − 5% = 95% = 0.95   percent to be paid

$276.40 × 0.95 = $262.58

Multiple Discounts

Sometimes, multiple discounts are applied. In that case, the first discount is applied to the list price. The second discount is applied to the net price after the first discount, with each additional discount applied to the previously calculated net price. Apply the discounts in the order listed.

Example: A contractor purchases framing lumber in quantity and receives discounts of 20% and 7%. If the material list price is $6,350, what is the net price?

$6,350 x 0.20 = $1,270 first discount amount

$6,350 − 1270 = $5,080 net after first discount

$5,080 x 0.07 = $355.60 second discount amount

$5,080 − 355.60 = $4,724.40 final net price

PRACTICAL PROBLEMS

For problems 1–3, find the amount of the discount at the given percentage and list price.

1.   12% discount on a purchase of $2,750.00   _____

2.   8% discount on a purchase of $254.00   _____

3.   25% discount on a purchase of $316.00   _____

For problems 4–6, find the net price after the specified discount is applied.

4.   22% discount on a purchase of $658.50   _____

5.   7% discount on a purchase of $21,585.00   _____

6.   14% discount on a purchase of $5,750.00   _____

For problems 7–10, find the amount of each discount and the net price after the discounts are applied.

7.   Discounts of 5% and 7% on a purchase of $1,850.00   _____

8.   Discounts of 10% and 5% on a purchase of $795.00   _____

9.   Discounts of 12%, 4%, and 2% on a purchase of $1,055.00   _____

10.   Discounts of 10%, 7%, and 5% on a purchase of $42,560.00   _____

11.   A computer supplier allows educators to purchase supplies and parts at a 15% discount. If the invoice is paid within 30 days, another 5% discount is allowed. A school ordered a disk drive with a list price of $279.00 and paid within the 30-day limit. Find the net price.   _____

12. Supplier A charges $6.25 for each ½" HSS drill bit, less 15%. Supplier B charges $5.80 for the same bit, with a 5% discount. What is the net cost from each supplier?

A _____

B _____

13. A model year clearance on appliances had discounts of 30% and 35%.

a. What is the net price for a range with a list price of $860?

_____

b. What is the net price expressed as a percent of the list price?

_____

14. The price of a CAD software package is listed at $495. If at least five packages are purchased, a 25% discount is applied. Since the purchaser is a school, another 10% discount is given. Find the cost of purchasing ten of the CAD software packages.

_____

15. Calculate the amount of each discount and the net price if discounts of 8%, 6%, and 3% are applied to a purchase of $1,000.00. Then, work the problem by finding the percent to be *paid* for each discount, multiplying the percents together, and then multiplying their product by the list price. Compare the two answers.

_____

16. An automotive repair shop priced a diagnostic testing machine from two sources. The list price was $5,800 with a 10% discount available from one supplier or $5,995 with discounts of 8% and 3% from another supplier.

a. Which is the lower net price?

_____

b. What is the difference in the net prices?

_____

!17. If a store advertises a sale price of $189.00 for a software program and the original list price is $249.00, what is the percent discount?

_____

18. Endmills ordered by an aircraft manufacturing plant are priced at $62.50 each for 1⅛" diameter, $58.00 for ¾" diameter, and $55.00 for ½" diameter. When at least 20 endmills of the same size are ordered, a 15% discount is applied to that item. If the total list cost of the order exceeds $1,000, a discount of 3% is applied to the total order. The aircraft company ordered thirty 1⅛" endmills, forty ¾" endmills, and eight ½" endmills. What is the net cost of the order?

_____

Unit 24 Averages and Estimates

BASIC PRINCIPLES OF AVERAGES

Averages are used often in business and industry. An average is found by adding a series of values and dividing their sum by the number of values added. For example, if five numbers are added, their sum is divided by five.

Example: If the width of a flange is measured on four randomly selected parts, with the dimensions recorded as listed below, find the average size.

> Part A: 3.428"   Part B: 3.426"
>
> Part C: 3.431"   Part D: 3.435"

> 3.428"
> 3.426"
> 3.431"
> 3.435"
> 13.720"   13.720" / 4 = 3.430"

BASIC PRINCIPLES OF ESTIMATES

Estimating is an important skill for industrial technologists. The term "estimate" has two meanings in technology. In math texts, estimation is defined as a way to determine an approximate value and can also be used to check the reasonableness of a calculated value. Good estimation skills are necessary to determine whether an answer obtained by calculation is reasonable. In industrial technology, the estimation can also refer to the process used to predict the cost of a project. Part of the necessary skills in estimation involve experience, such as knowing how long it should take to perform a task. Both meanings of estimation will be seen in this unit.

Example: A student used a calculator to find the product of 28.50 and 1,050. The display on the calculator was 2,992,500. Does the answer appear to be reasonable?

Mentally rounding the values to 30 and 1,000 and multiplying would yield a product of 30×10^3 or 30,000. It appears that perhaps the decimal point was not entered correctly. Recalculating the problem yielded the value 29,925, which appears to be reasonable.

PRACTICAL PROBLEMS

For problems 1–5, use rounding to estimate the answer. Then calculate the actual value and find the difference between the estimated and actual values.

1. 12 × 97 _____

2. 810 × 90 _____

3. 2,250 × 37 _____

4. 320 × 28 _____

5. 3,200 × 1,900 _____

For problems 6–9, find the average of the values to the same precision as the values.

6. 375.5   428.3   358.9   465.7 _____

7. 18,250   16,875   14,864 _____

8. 2,186   2,254   2,188   1,956   2,010 _____

9. 42.55   41.83   39.87   41.71 _____

10. Estimate the average using rounded values, and find the percent error (error/actual value).

    2,100   4,150   3,750   1,940 _____

11. The first five aircraft wing supports produced required 31.5, 33.8, 29.7, 30.9, and 28.2 hours respectively. What is the average production time for a part, to the nearest tenth of an hour? _____

12. The cost of lathe bits is $3.95 each. Use mental math or rounding to estimate the cost of 12 bits. _____

13. A carpenter estimates that it takes 1¾ hours to install a window.

    a. Use rounding to obtain an estimated time to install 12 windows. _____

    b. If he charges $18 per hour, what is the estimated cost of the project? (Note: If both numbers need to be rounded, the estimate will be closer to the actual value if one of the values is rounded "up" and the other is rounded "down.") _____

14. A student uses a modem to access research materials on a computer service.

   a. If his daily on-line times for 1 week were 22 minutes, 17 minutes, 8 minutes, 49 minutes, and 15 minutes, what is his average daily time on-line to the nearest minute?   _____

   b. If the cost is $0.08 per minute, what is his average daily cost?   _____

   c. What is his estimated monthly cost, using 30 days per month?   _____

15. A contractor allows a customer $18 per square yard for floor covering, and estimated 205 square yards for a house under construction.

   a. What is the contractor's allowance for floor covering?   _____

   b. If the customer selected carpet at $21.50 per yard and sheet vinyl at $16.00 per yard and the house required 65 square yards of vinyl and 130 square yards of carpet, what was the actual cost?   _____

   c. Express the difference as a percentage of the contractor's estimate to the nearest percent.   _____

16. A machine shop produced 6 sample parts for a customer. The machining times required were 1.2 hours, 1.5 hours, 1.1 hours, 0.9 hours, 1.0 hour, and 1.2 hours.

   a. What is the average machining time to the nearest tenth of an hour?   _____

   b. If the average time is used to prepare a bid for 120 more parts at an hourly rate of $24.50, what is the total amount of the bid?   _____

17. The mean (average) time between failures (MTBF) is often used as a way of measuring the reliability of a product. A printer company tested five printers, with the run times before failure recorded as 4,120 hours, 7,200 hours, 3,508 hours, 5,125 hours, and 7,984 hours. What is the MTBF for those five printers to the nearest hundred hours?   _____

18. An employee submitted a suggestion to his company which is estimated to save 2.3 minutes per assembly operation on an automobile body. Use rounding to estimate annual time savings if the factory produces an average of 9,000 of that model per year.   _____

19. A free-lance drafter for an engine manufacturing company estimates that a typical exploded view drawing for an owner's manual requires 20 hours of work. After completing 6 drawings for the company, she decided to check the accuracy of her estimate. If the times required for each of the drawings completed were 22.5 hours, 16.75 hours, 19 hours, 14.25 hours, 25.75 hours, and 32 hours, what is the percentage error (error ÷ actual average) in her estimate?　　_____

!20. The life of a color cartridge for an inkjet printer is estimated to be 200 pages at 15% print coverage per page. The department manager of a graphic arts company is trying to estimate the life of a cartridge for her department's typical print jobs. The first 8 print jobs done using the printer were estimated to have the following percentages of print coverage on each page: 25%, 10%, 7%, 10%, 12%, 6%, 20%, and 10%. Using the average of the first 8 jobs, what is the expected life of the cartridge?　　_____

Unit 25 Review of Problems Involving
Percents, Averages, and Estimates

PRACTICAL PROBLEMS

1. Convert $\frac{5}{16}$ to percent.   _____

2. Convert 87.5% to a fraction.   _____

3. Convert 175% to a mixed number.   _____

4. 288 is what percent of 640?   _____

5. What number is 8% of 3,250?   _____

6. 210 is 35% of what number?   _____

7. Calculate the simple interest and final payoff amount for a loan of
    $2,200, with an annual interest rate of 6% for 2 years.   _____

8. Find the monthly interest rate for a loan with an annual interest rate of 8
    percent.   _____

9. What is the net price for a drill if its list price is $89.95 and a 15%
    discount is offered?   _____

10. Find the average of 84.52, 78.35, 82.11, and 89.62.   _____

11. The estimated manufacturing cost for a door panel was $52.68. If the
    actual cost was $54.26, express the cost overrun as a percentage of the
    estimated cost, to the nearest percent.   _____

12. According to a company policy, the acceptable reject level for an
    injection-molded plastic part is 1.5%. For a production run of 36,400
    parts, what is the maximum allowable number of rejected parts?   _____

13. A company invested $32,000 in a new technology project which
    produced an annual rate of return on investment of 5.85%. How much
    was earned on the investment?   _____

14. A rust-resistant primer used under the finish coat on automobile parts is available from two suppliers. Supplier A charges $328 for a 55-gallon drum, but allows a 15% discount. Supplier B charges $319.50 for the same quantity, and allows a 10% discount. Which is the lower net price? _____

15. The percent elongation is used as a measure of the ductility of a material. Testing involves applying a tensile load to a material and expressing the change in length of the sample (before and after the test) as a percentage of the original length. Using l_F = final length and l_0 = original length, the formula is stated as $\% \text{elongation} = \frac{l_F - l_0}{l_0} \times 100$.

Find the % elongation if $l_F = 2.185"$ and $l_0 = 2.000"$. _____

16. In the AISI and SAE four-digit steel designation system, the last 2 digits give the carbon content in *hundredths of a percent*. What is the percent carbon in 4620 steel? Express the carbon content in decimal form. _____

17. If low-density polyethylene has a tensile strength of 3,000 psi and high-density polyethylene has a tensile strength of 5,500 psi, express the strength of LD polyethylene as a percent of the strength of HD polyethylene. (Round to the nearest percent.) _____

18. An architecture firm borrowed $38,500 for 1 year at 9% interest to purchase a CAD system network.

   a. What is the amount of interest due at the end of the year? _____

   b. What would be the total amount due if the loan is extended for another six months? _____

19. A drafting supply company pays shipping charges if an order totals more than $1,000. Otherwise, a shipping/handling charge of 5% of the order value is added. If a vocational school ordered 10 drafting compass sets at $54.00 each, twelve sets of scales at $4.80 each, 3 packages of 18"x24" paper at $82.00 each, and 2 dozen plotter pens at $15.60 per dozen, what is the total cost of the order, including shipping if applicable? _____

20. The time required to apply a reflective tape to a helicopter support bracket was recorded on the first six parts produced. The times required were 32, 38, 29, 31, 32, and 28 seconds respectively. What is the average installation time for the tape? _____

21. Voltage readings were measured on a circuit supplying a mainframe computer unit for five days. The readings were recorded as: 117 volts, 112 volts, 128 volts, 121 volts, and 120 volts.

    a. What is the average voltage? _____

    b. If the voltage specifications for the machine state that the voltage should not vary more than 5% from the desired voltage of 120 volts, are any of the daily readings out of specification? _____

22. A graphic artist is preparing a diagram for a technical manual. The bar shown below is to be used to illustrate the percentage of carbon in various types of steel and cast iron. Percentages from 0 to 2% are to be shown on a bar which is 3.5" long. Calculate the boundary points between segments to represent the types of steel or cast iron, using the data in the chart.

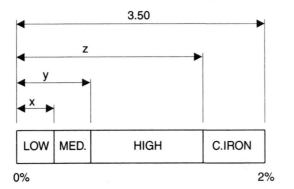

Percent carbon	Material type	Upper boundary
< 0.3%	Low carbon steel	
0.3 – 0.6%	Medium carbon steel	
0.6 – 1.5%	High carbon steel	
> 1.5%	Cast iron	

!23. Before using a tap to produce internal threads, a hole is drilled using a tap drill. Recommended tap drills for a variety of NC (national coarse) threads are shown in the chart below. For each nominal thread size and each fractional tap drill, find its decimal equivalent to 3 decimal places. For each thread, divide the tap drill size by the nominal thread size to find the % full thread, and convert each to percent form. Based on the values obtained, what appears to be a common % full thread value?   _____

Nominal thread	Threads per inch	Decimal equiv.	Tap drill size	Decimal equiv.	% full thread
$\frac{1}{4}$	20		#7	.2010	
$\frac{5}{16}$	18		F	.2570	
$\frac{7}{16}$	14		U	.3680	
$\frac{1}{2}$	13		$\frac{27}{64}$		
$\frac{9}{16}$	12		$\frac{31}{64}$		
$\frac{3}{4}$	10		$\frac{21}{32}$		

Exponents and Roots

Exponents are an important concept in mathematics. Powers of 10 form the basis for our number system. Powers of 2 are the basis for the binary number system used by computers. Many physical properties in the world can be expressed using exponential notation. (Remember that mathematics is a "language" used to describe phenomena and properties encountered in the world.) In this section, a variety of uses of exponents will be practiced.

Roots are also important and are closely related to powers. The term "root" is used in several different ways in mathematics and engineering. The most common meaning (the one used in this section) is a number which is raised to a power to yield a particular number, as in square roots and cube roots. (Note: In engineering and mathematics texts, the term "root" is often used to mean the solution to an equation.)

It is important to understand the relationship between roots and exponents. Also, *learn how to use a scientific calculator to calculate expressions containing roots and exponents.* While they were often calculated manually in the past, the process has been made much simpler by the use of new technologies, such as calculators or computers.

Unit 26 Exponents and Order of Operations with Exponents

BASIC PRINCIPLES OF EXPONENTS

An *exponent* is a math symbol used to indicate that a number is multiplied by itself a specified number of times. An exponent is also referred to as the *power* to which a number is raised. It is written as a smaller number above and to the immediate right of the number being multiplied, as in 2^3 or 4^2.

Numbers which are multiplied together are called *factors*. In the case of powers, where the same number is being multiplied by itself, it is also called the *base*. Frequently used powers are given specific names:

A number raised to the power 2 is said to be *squared*.

A number raised to the power 3 is said to be *cubed*.

By definition, in order to follow established rules for working with exponents, when any number is raised to the zero power, the value is one (1). Also, remember that a negative exponent represents the reciprocal of the value.

The order of operations introduced in Unit 5 should be followed when evaluating an expression containing exponents. Notation involving exponents should be written so that the base value is clear. If a single digit or variable is raised to a power, no parentheses are needed. However, if the expression to be raised to a power has more than one factor or term, the base must be written in parentheses.

Examples:

$$2^3 = 2 \times 2 \times 2 = 8$$
$$(2x)^3 = (2x)(2x)(2x) = 2^3\, x^3 = 8x^3$$
$$3^{-2} = \frac{1}{3^2} = \frac{1}{9}$$
$$4^2 \times 2^{-2} + 27 = 16 \times \frac{1}{2^2} + 27 = 16 \times \frac{1}{4} + 27 = 4 + 27 = 31$$

POWERS OF 10 AND WHOLE NUMBER PLACE VALUE

The decimal number system is also referred to as a *base 10* system because it is based on 10 digits and powers with 10 as the base. In Section 1, each digit position in a whole number was identified as having a place value. The actual value represented by each digit is the digit multiplied by the correct power of 10. The number 384 can be written in expanded notation as

$$384 = (3 \times 100) + (8 \times 10) + (4 \times 1) = (3 \times 10^2) + (8 \times 10^1) + (4 \times 10^0)$$

Some of the powers of ten representing each place value are shown below.

$$\ldots 10^3 \qquad 10^2 \qquad 10^1 \quad 10^0 \quad 10^{-1} \qquad 10^{-2} \ldots$$
$$\ldots \text{thousands} \quad \text{hundreds} \quad \text{tens} \quad \text{ones} \quad \text{tenths} \quad \text{hundredths} \ldots$$

The use of powers of 10 was also illustrated in scientific notation. Multiplying or dividing by powers of 10 has the effect of moving the decimal place.

BASIC PRINCIPLES OF BINARY NUMBERS

The binary numbers are a system used in the computer industry. The number system behaves very similarly to the decimal (base 10) system, except that it has the number 2 used as the base which is raised to various powers. In the illustration below, notice the similarity to the decimal (base 10) system illustrated above.

$$\ldots 2^3 \qquad 2^2 \qquad 2^1 \quad 2^0 \quad 2^{-1} \qquad 2^{-2} \ldots$$
$$\ldots \text{eights} \quad \text{fours} \quad \text{twos} \quad \text{ones} \quad \text{halves} \quad \text{fourths} \ldots$$

Note that the words (eights, fours, twos, etc.) are expressed in decimal terms so that their meaning is familiar. All binary numbers are written as a series of 0's and 1's, since the system has only those two digits.

Example: Convert 1101_2 to a base 10 number.

$$(1 \times 2^3) + (1 \times 2^2) + (0 \times 2^1) + (1 \times 2^0)$$
$$= (1 \times 8) + (1 \times 4) + (0 \times 2) + (1 \times 1) = 13_{10}$$

 ## CALCULATOR USE

Several keys on scientific calculators are used to enter exponents, including the $[x^2]$ and $[y^x]$ keys. To multiply a number by itself (square the number), enter the number and then press the $[x^2]$ key. On most calculators, the calculation is immediate, without having to press the equal key [=]. To raise a number to a power other than 2, enter the number, press the $[y^x]$ key, enter the exponent, and press the equal key [=]. You may need to refer to your calculator manual to determine how the keys are marked and the order in which the numbers should be entered.

Also, remember that the [Exp] or $[10^x]$ keys can be used to enter powers of 10 if your calculator has either or both of those keys.

PRACTICAL PROBLEMS

For questions 1–8, evaluate each of the following expressions:

1. 5^3 _____

2. 2^5 _____

3. $3^3 + 2^4 - 5^2$ _____

4. $(3 + 2^3) + (5^3 - 4^2)$ _____

5. $(3^3 + 4^2)^2$ _____

6. $2^6 - 3^3$ _____

7. $25^2 - 8^3$ _____

8. $10^4 - 10^3 + 10^2$ _____

9. The resistivity of copper is 1.724×10^{-6} ohm-centimeters. Express the resistivity in decimal form. _____

10. The area of a square can be calculated by multiplying the length of the side by itself. For a square 8" on a side, find its area. (Be sure to include the units as square inches.) _____

11. The power for a circuit can be expressed as $P = I^2 R$. If a circuit is carrying 2.1 amps (I) and its resistance is 15 ohms (R), find P in watts. _____

12. The kinetic energy of an object is calculated using the formula $KE = (0.5)mv^2$, where m is the mass of the body and v is its velocity.

    a. For an object with a mass of 1,350 kg traveling at 22 meters per second, what is its kinetic energy? (Include the unit "joules" with the value.) _____

    b. If its velocity doubles to 44 meters per second, what is its kinetic energy? _____

    c. What is the ratio of kinetic energy at the faster velocity to kinetic energy at the lower speed? _____

13. The cross-sectional area of a wire can be found using the formula for area of a circle, which is stated as $A = \pi r^2$, where A is the area, π is approximated by 3.1416, and r is the radius.

   a. For a steel wire of diameter 0.102", find its cross-sectional area to the nearest ten-thousandth. _____

   b. For a steel wire with a 0.204" diameter, find the cross-sectional area. (Remember that radius is ½ of diameter.) _____

   c. Write a ratio of the area of the larger wire to the area of the smaller wire. _____

   d. Complete the statement: When the diameter of a wire doubles, the cross-sectional area is increased by a factor of _____. _____

14. Convert 10111_2 to a base 10 number. _____

15. The power radiated by a heating element is given by the formula P $= I^2R$, as described in problem 11 above. If the heating element has a resistance of 23.1 ohms and carries 10.8 amps, find P, rounded to the nearest watt. _____

16. The wattage (W) in a circuit can be calculated by multiplying the square of the current (I), in amperes, by the resistance (R), in ohms. Stated in mathematical symbols, $W = I^2 R$. If a circuit has a resistance of 6.25 ohms and a current of 1.2 amps, what is the wattage? _____

17. Convert 1110101_2 to a base 10 number. _____

18. Using the wattage formula $P = I^2 R$, find the wattage for a circuit with a resistance of 13.7 ohms and a current of 8.75 amps, rounded to the nearest amp. _____

19. Convert 1101011_2 to a base 10 number. _____

20. While testing a new polymer material for catchers' masks, a test is done to determine the kinetic energy of a baseball traveling at various velocities. One trial resulted in a velocity of 43 meters per second, with a ball which had a mass of 0.22 kilogram. Using the formula for kinetic energy, $KE = (0.5)mv^2$, find the kinetic energy. _____

!21. Another common number system used in the computer industry is the *hexadecimal* system. Numbers in the hexadecimal system are indicated by writing a subscript h or 16, as in $3E8_h$ or $4F_{16}$. It has 16 digits: 0, 1, 2, 3, 4, 5, 6, 7, 8, 9, A, B, C, D, E and F, which are equivalent to the numbers 0 through 15 in the base 10 system. The hexadecimal system allows larger values to be expressed with fewer digits per number than base 10 would require.

Convert each hexadecimal number below to the base 10 (decimal) system using the process described in the Basic Principles of Binary Numbers section above, substituting base 16 for base 2 in the example.

a. 82E   _____

b. 24CD   _____

c. D000   _____

d. 2F8   _____

Unit 27 Roots

BASIC PRINCIPLES OF ROOTS

The topic of roots is closely related to the topic of powers. A *root* is a number which, when raised to a specific power, results in the specified value.

The square root of 16 is 4 because $4^2 = 16$.

The cube root of 8 is 2 because $2^3 = 8$.

The fourth root of 81 is 3 because $3^4 = 81$.

A *radical sign* ($\sqrt{\ }$) is used to indicate a root. If the root is a square root, no *index* is needed. Otherwise, a 3 is used to indicate a cube root, 4 is used to indicate a fourth root, etc. In the expression below, read as the "nth root of x", the n is the index, and the x is the *radicand* (the number under the radical sign).

$$\sqrt[n]{x}$$

(Note: You may also see roots written as fractional exponents, with a numerator of 1 and the index used as the denominator, in math texts. For example, the cube root of 8 may be written as $8^{1/3} = 2$.)

$$\sqrt{16} = 4 \qquad \sqrt[3]{8} = 2 \qquad \sqrt[4]{81} = 3$$

Although a number may have both a positive and negative root, the positive root is called the *principal root* and is the root most commonly applicable to practical problems. As an example, 5^2 and $(-5)^2$ both equal 25, but only positive 5 has meaning as the length of the side of a square whose area is 25.

There are several ways to determine a root. In years past, roots were often calculated manually or by looking in tables. Currently, it is expected that most students will use a scientific calculator if needed, due to their common availability and ease of use.

 CALCULATOR USE

Calculators which can calculate roots have keys which may be marked in several ways, including $[\sqrt{\ }]$, $[\sqrt[3]{\ }]$, or $[\sqrt[x]{\ }]$. The $[\sqrt{\ }]$ key is used to calculate a square root. On calculators which use *direct* algebraic logic, press the $[\sqrt{\ }]$ key, enter the radicand, and press the [=] key. For calculators using algebraic logic, enter the radicand and press the $[\sqrt{\ }]$ key. The square root will normally be displayed immediately, without having to press the [=] key.

The $[\sqrt[3]{}]$ key is used to calculate a cube root. On calculators which use direct algebraic logic, press the $[\sqrt[3]{}]$ key, enter the radicand, and press the [=] key. For calculators which use algebraic logic, enter the radicand first and then press the $[\sqrt[3]{}]$ key to display the root.

The $[\sqrt[x]{}]$ key can be used to find roots other than square or cube roots. For direct algebraic logic, enter the index first, then press the $[\sqrt[x]{}]$ key, then enter the radicand and press the [=] key. A similar series of keystrokes is used on most algebraic calculators. (Note: There is considerable variety among different manufacturers in the way that keys are labeled and the order of keystrokes which are used to calculate roots. *It is especially important to consult your owner's manual if you plan to use your calculator to calculate roots.*)

PRACTICAL PROBLEMS

For questions 1–13, evaluate the expression. Express your answer to the nearest tenth, unless otherwise specified.

1. $\sqrt{64}$ _____

2. $\sqrt{729}$ _____

3. $\sqrt{273}$ _____

4. $\sqrt[3]{2,744}$ _____

5. $\sqrt{625} - \sqrt{256}$ _____

6. $\sqrt[4]{256}$ _____

7. $\sqrt{6^2 + 8^2}$ _____

8. $\sqrt[3]{95}$ _____

9. $\sqrt[3]{\dfrac{2,800}{25}}$ _____

10. $\sqrt[4]{820 \times 6}$ _____

11. $\sqrt{\dfrac{2,600}{24}}$ to the nearest hundredth   _____

12. $\sqrt{1{,}050 \times 24}$ to the nearest hundredth

                                                  _____

13. $\sqrt{\dfrac{628}{(4)(3.14)}}$ to the nearest tenth

     _____

14.   A cylindrical tank has a volume (V) of 235 cubic feet. If the height (H) is 6', find the radius (R).

$$R = \sqrt{\dfrac{V}{H\pi}}$$

     _____

15.   An electric range heating element uses 2,700 watts of power (P). The resistance (R) of the device is 23.1 ohms. On what voltage (E) does the heating element operate?

$$E = \sqrt{P \times R}$$

     _____

16.   A contractor ordered 1,355 square feet of vinyl tile for a square room. To the next whole foot, what is the length of one side of the room? Use the formula $s = \sqrt{area}$ where s is the length of a side and area is the number of square feet in the room.

     _____

17.   The top of a circular storage building has an area of 490 square feet. The radius can be calculated using the expression $r = \sqrt{\dfrac{area}{3.14}}$. Using 3.14 as an approximate value for π, find the radius of the building.

     _____

18.   The kinetic energy of an object is 225 J. If its mass (m) is 50 kg, find its velocity using the expression $v = \sqrt{\dfrac{2 \times KE}{m}}$. Include the unit m/s as part of your answer. (Note: The derivation of the correct units is given as an example of Dimensional Analysis in Section 8.)

     _____

Unit 28 Review and Combined Operations on Exponents and Roots

PRACTICAL PROBLEMS

For questions 1–8, evaluate the expression. Round to hundredths if rounding is needed.

1. $(3^2 + 4^3)^2$　　　　　　　　　　　　　　_____

2. $10^5 + 10^3 - 10^2$　　　　　　　　　　　_____

3. $2^5 - 3^2$　　　　　　　　　　　　　　　_____

4. $\sqrt{232.6}$　　　　　　　　　　　　　　_____

5. $\sqrt{12^2 + 5^2}$　　　　　　　　　　　　_____

6. $\sqrt[3]{30}$　　　　　　　　　　　　　　　_____

7. $\sqrt{3^2 + 4^2}$　　　　　　　　　　　　_____

8. $\sqrt{1,500 \times 7.4}$　　　　　　　　　　_____

9. $\sqrt{13^2}$　　　　　　　　　　　　　　_____

10. $(\sqrt[3]{1,728})^3$　　　　　　　　　　　_____

11. Convert 1001111_2 to base 10.　　　　　_____

12. For a circuit with 3.8 amps and a resistance of 12 ohms, find P (using P= I^2R) to the nearest watt.　　　　　_____

13. Find the length of the side of a square whose area is 191.8 in^2 to the nearest thousandth of an inch.　　　　_____

14. A wire has a cross-sectional area (a) of 0.049 in^2. Find its radius to the nearest thousandth of an inch, using 3.14 for π and the formula $r = \sqrt{\dfrac{a}{\pi}}$.　　_____

15. Find the kinetic energy of a moving object with mass 1.3 kilograms and a velocity of 50 meters per second. Use the formula KE = $(0.5)mv^2$.   _____

16. A cylindrical tank has a volume (V) of 325 cubic feet. If the height (H) is 5 feet, find the radius (R).   _____

$$R = \sqrt{\frac{V}{H\pi}}$$

17. Find the volume of a tank (V) with a radius of 3.5 meters (R) and a height (H) of 4.8 meters. Use the formula V = $\pi R^2 H$.   _____

Measurement

BASIC PRINCIPLES OF MEASUREMENT

Measurement is an important part of technical and engineering work. Two main systems of measurement are in use in the world today, the SI (metric) system and the U.S. Customary system. Both systems will be used in the practical problems; charts showing conversion factors are provided in the Appendix.

CONVERSION FACTORS USED IN MEASUREMENTS

Measurements can be changed from one unit to another unit in the same system or to units in another measurement system.

Converting between Units in the Metric System

When converting from one unit to another in the metric system, it is possible to convert by just shifting the decimal to the right the correct number of places when changing to smaller units, or to the left when changing to larger units. In order to shift the decimal, you must understand the relationship between various metric prefixes such as *kilo, mega, centi,* etc. The chart below lists some common prefixes used in engineering and technical applications. When converting from smaller units to larger units, the number of units will decrease. When converting from larger units to smaller units, the number of units will increase.

Example: 1,000 mm = 1 m   1 km = 1,000 m

Prefixes Used in Metric System

Prefix	Symbol	Prefix Value	Prefix	Symbol	Prefix Value
giga	G	10^9	deci	d	10^{-1}
mega	M	10^6	centi	c	10^{-2}
kilo	k	10^3	milli	m	10^{-3}
hecto	h	10^2	micro	μ	10^{-6}
deka	da	10^1	nano	n	10^{-9}

Converting Between Units in the U.S. System or Between Systems

The relationship between the two units must be known in order to write an appropriate conversion factor. (A table of unit conversions is included in the Appendix.) If the relationship is stated as "1 inch is approximately 2.54 cm," two conversion factors can be written:

$$\frac{1''}{2.54 \ cm}; \qquad \frac{2.54 \ cm}{1''}$$

Use the conversion factor form which allows the units to be canceled as needed. For example, if the problem starts with inches in the numerator and another unit is needed, select a conversion factor with inches in the denominator and then divide out the inch units. The use of the conversion factors does not change the value of the expression, because it basically involves multiplying by the value "1".

Example: Convert 3.50" to centimeters.

$$\frac{3.50''}{1} \times \frac{2.54 \ cm}{1''} = 8.89 \ cm$$

Example: Convert 4 gallons to liters.

$$\frac{4 \ gallons}{1} \times \frac{3.785 \ L}{1 \ gallon} = 15.14 \ L$$

BASIC PRINCIPLES OF DIMENSIONAL ANALYSIS

Units are considered carefully in *dimensional analysis*. In that process, units are handled according to algebraic principles. Units may be multiplied, divided ("canceled out"), or expressed in a different form using conversion factors. In some problems, several unit conversions may be needed. For example, if the unit J (joules) is used, you may need to substitute N-m for J before continuing to work with the units. The conversion shown below illustrates how the units for Problem 18 in Unit 27 were derived.

$$\sqrt{\frac{J}{kg}} = \sqrt{\frac{Nm}{kg}} = \sqrt{\frac{\frac{kgm}{s^2} \times \frac{m}{1}}{kg}} = \sqrt{\frac{m^2}{s^2}} = \frac{m}{s}$$

One important use of dimensional analysis is in verifying that a problem has been expressed or written correctly. The problem can be "worked" using only the units before the numbers are considered. If the correct units are achieved after simplification, the problem can be assumed to be set up correctly.

Example of a correctly stated conversion: $\dfrac{30 \text{ mi}}{1 \text{ hr}} \times \dfrac{1 \text{ hr}}{60 \text{ min}} \times \dfrac{1 \text{ min}}{60 \text{ s}} \times \dfrac{5{,}280 \text{ ft}}{1 \text{ mi}} = \dfrac{44 \text{ ft}}{\text{s}}$

Example of an incorrectly stated conversion: $\dfrac{30 \text{ mi}}{1 \text{ hr}} \times \dfrac{60 \text{ min}}{1 \text{ hr}} \times \dfrac{60 \text{ s}}{1 \text{ min}} \times \dfrac{5{,}280 \text{ ft}}{1 \text{ mi}} = \dfrac{\text{s} \times \text{ft}}{\text{hr} \times \text{hr}}$

When working with measurements, be sure to write the units as well as the number.

Unit 29 Length and Angle Measurement

LENGTH MEASUREMENT IN THE U.S. CUSTOMARY SYSTEM

Units of length in the U.S. system are based on inches, feet, yards, miles, etc. The relationships between the various units are shown in the chart below. Each of the relationships can be written as a conversion factor using the procedure described in the Section 8 introduction.

U.S. Length unit	. . . is equivalent to:
1 foot	12 inches
1 yard	3 feet
1 mile	5,280 feet
1 mile	1,760 yards

Measurement in the U.S. system is often expressed as a series of mixed units, as in 2 feet 4 inches (or 2' 4"). This is common in the construction and architectural drafting fields. Addition and subtraction of that type of values may be handled by writing the numbers in columns with like units in each column. In the case of subtraction, regrouping may be needed. When regrouping from feet to inches, for example, 1' may be rewritten as 12" and the 12" added to the inches in the original number. After completing the operation, the measurement should be written in the largest possible units.

Examples:

$$
\begin{array}{r}
3'\ 6" \\
+\ 4'\ 8" \\
\hline
7'\ 14"
\end{array}
\quad\Rightarrow\quad 7'\ +\ 1'2"\ =\ 8'2"
$$

$$
\begin{array}{r}
8'\ 6" \\
-\ 4'\ 9" \\
\hline
\end{array}
\quad=\quad
\begin{array}{r}
7'\ 18" \\
-\ 4'\ 9" \\
\hline
3'\ 9"
\end{array}
$$

Be sure to add only like units, such as feet + feet.

LENGTH MEASUREMENT IN THE SI (METRIC) SYSTEM

The standard unit for length measurement in the SI system is the meter. Other metric system measurements are "power of 10" multiples of the meter. Some of the metric prefixes commonly used in technical and engineering work were listed in Section 8.

The conversion of metric units can be done using conversion factors. Use a factor which has 1 as the numeric value for the larger unit to avoid decimal division.

Example: $\dfrac{1\ m}{1{,}000\ mm}$ vs. $\dfrac{.001\ mm}{1\ mm}$

Example: Convert 375 mm to meters.

$$\frac{375\ mm}{1} \times \frac{1\ m}{1{,}000\ mm} = 0.375\ m$$

An alternate method for converting between metric units is to move the decimal point the correct number of places to the right (a larger number) when changing to smaller units or to the left (a smaller number) when changing to larger units.

Example: Convert 375 mm to meters.

$$375\ mm = 0.375\ m$$

Since 1 mm = $\frac{1}{1000}$ (or 1000 mm = 10^3 mm = 1 m), the decimal point should be moved 3 places. Meters are larger units, so the number of units required for the same length is smaller and the decimal point is moved to the left.

CONVERSION OF LENGTH MEASUREMENTS BETWEEN SYSTEMS

Conversion factors needed to convert from the U.S. system to the metric system or vice versa can be written from relationships in the table below.

U.S. Customary unit	SI (metric) equivalent
1 inch (in)	2.54 cm
1 foot (ft)	0.3048 m
1 yard (yd)	0.9144 m
1 mile (mi)	1,609 km

Example: Convert 65 miles to kilometers.

$$\frac{65\ mi}{1} \times \frac{1.609\ km}{1\ mi} = 104.585\ km$$

ANGLE MEASUREMENT

Two common units of measurement for angles are degrees and radians. In the degree system, a complete circle contains 360 degrees (360°). Using the radian system, a complete circle contains 2π radians. The degree system is often used in manufacturing and construction to describe angles on parts. In many engineering calculations, radians are used. Standard position for measuring angles assumes that the initial side of the angle is horizontal, with the vertex of the angle on the left, and the angle measured counterclockwise. This reference system is used in many CAD systems and for the unit circle used in trigonometry, and is illustrated below.

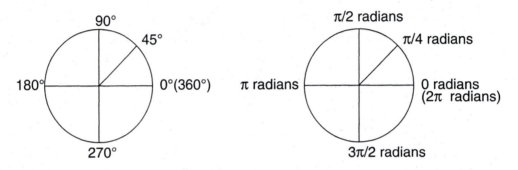

Converting between Degrees and Radians

Using the relationship 360° = 2π radians, two conversion factors can be written:

$$\frac{360°}{2 \pi \ radians} \qquad \frac{2 \pi \ radians}{360°} .$$

Conversions can be done by multiplying the original measurement by the appropriate conversion factor.

Example: Convert 45° to radians.

$$45° \times \frac{2 \pi \ radians}{360°} = \frac{\pi}{4} \ radians$$

 ## CALCULATOR USE

Most scientific calculators are able to handle angle measurements in both degrees and radians. Normally, a "mode" must be set and then angle measurements are handled in that system until it is changed. Some calculators also will convert from one angle system to the other. See the owner's manual for specific instructions on how to set the mode on your calculator, as there is considerable variation between brands. See problem 25 for a method of using a standard calculator to add feet and inches.

PRACTICAL PROBLEMS

For problems 1–9, use a conversion factor and dimensional analysis to perform the conversions.

1. Convert 2.5 m to cm.   _____

2. Convert 3,805 mm to m.   _____

3. Convert 42.5 cm to mm.   _____

4. Convert 72 degrees to radians.   _____

5. Convert 240 degrees to radians.   _____

6. Convert 3π/4 radians to degrees.   _____

7. Convert 5π/4 radians to degrees.   _____

8. 7,392 feet is equivalent to how many miles?   _____

9. Express 27 miles to the nearest tenth of a kilometer.   _____

10. A bill of materials for a shelf unit lists 42 pieces of 1" square tubing 14" long, 6 pieces 5 feet long, and 10 pieces 27 inches long.

    a. What is the total length of steel tubing needed, neglecting the length used in each cut?   _____

    b. If the inventory has 9 pieces, each 12' long, is there enough in inventory to build the unit?   _____

11. A machine measures 48" long, 42" high, and 28" deep. A minimum clearance of 18" on the left end and 30" on the right end is required. If the machine is positioned with its length along the wall, what is the minimum wall space (in feet) required for the machine?   _____

12. In a manufacturing process, pieces 3" long are to be cut from round bar stock. Allowing ⅛" per cut, how many pieces can be cut from a piece 4' long? (Hint: a sketch may help.)   _____

13. The speed of sound at 0°C is 1,087 ft/s.

    a. Convert the speed of sound to miles per hour.   _____

    b. Express Mach 2 (twice the speed of sound) in meters/second.   _____

14. A part to be machined is shown below, dimensioned in decimal inches. Find each dimension to the nearest millimeter. What is the total length in millimeters?   L = _____

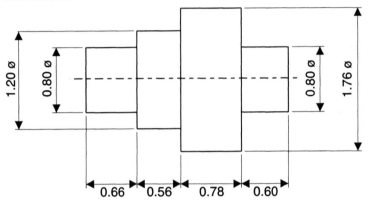 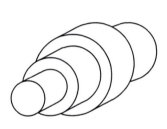

15. Calculate the metric measurements and complete the blanks in the statement, to the nearest tenth of a mm:
    The usual allowance for machine reaming is 1/64" (____mm) for holes under ½" (____mm) and 1/32" (____mm) for holes ½" and over.

16. A company must transport a load of finished parts to a city 185 miles from the factory. What is the distance in km?   _____

17. A scale normally used for floor plans on residential buildings is ¼" = 1'0".

    a. Write the conversion factors expressing the relationship between drawing size and actual size.   _____

    b. A house which is 86' long would be drawn how long?   _____

    c. A width of 42' would be drawn how long?   _____

18. For each scale illustration, write the length dimension shown.

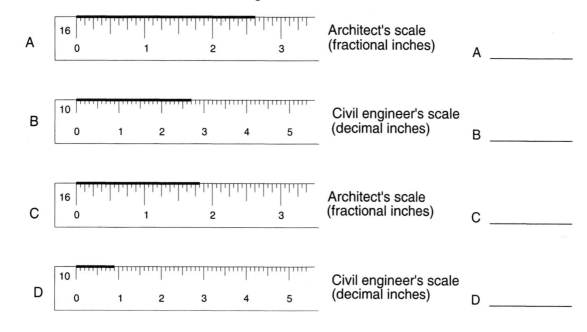

A   Architect's scale (fractional inches)   A _____

B   Civil engineer's scale (decimal inches)   B _____

C   Architect's scale (fractional inches)   C _____

D   Civil engineer's scale (decimal inches)   D _____

19. The length of a bandsaw blade is found as shown in the illustration. For the given dimensions, find the length of the blade needed, to the nearest inch. (The circumference of a circle is given by the formula $C = 2\pi r$, where π can be approximated by 3.1416 and r = radius of the circle.)   _____

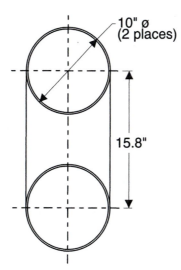

10" ø
(2 places)

15.8"

20. Measure the lines shown using the "16" scale on the architect's scale. Be sure that the zero (0) is aligned exactly with one end of the line, and read the measurement from the other end.

A _____

B _____

C _____

D _____

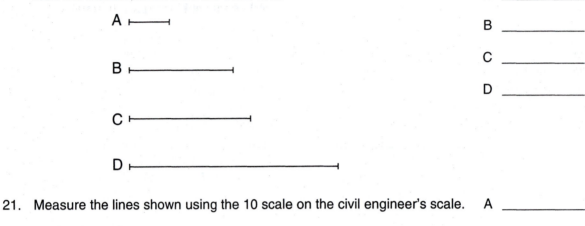

21. Measure the lines shown using the 10 scale on the civil engineer's scale.

A _____

B _____

C _____

D _____

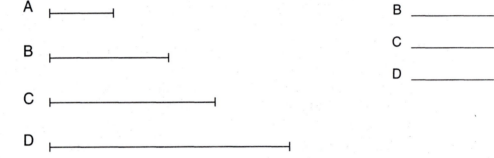

!22. A micrometer is a measurement instrument which uses a screw thread with 40 threads per inch. Therefore, one revolution of the spindle causes a change of ¼₀" in the measurement. For 24 revolutions of the spindle, what is the linear movement which would occur?

_____

!23. The 40 scale on the civil engineer's scale has increments of ¼₀". What is the decimal equivalent of ¼₀" inch? What distance would be represented by 5 of those increments? What is the decimal equivalent of ⅛"? Write a brief statement explaining how the 40 scale could be used to measure eighths of inches if a fractional scale were not available.

_____

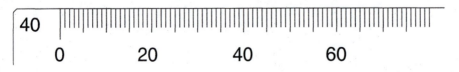

_____

_____

!24. The 50 scale on the civil engineer's scale has increments of ⅟₅₀". What is the decimal equivalent of ⅟₅₀"? Write a brief statement explaining how the 50 scale can be used to read hundredths of an inch, and the level of precision that can be obtained.

_____

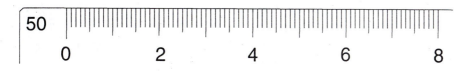

!25. A standard scientific calculator can be used to add measurements including both feet and inches. In order to use this process, each measurement must be entered as a decimal, with the feet value entered as the whole number part and the inch value entered as the decimal part. There should be enough zeroes in front of the decimal part to prevent the sum from carrying over to the whole number part. As an example, 27 feet 6 inches can be entered as 27.006 or 27.0006.

Using this process, add 24 feet 9 inches + 14 feet 7 inches + 8 feet 6 inches. Write the problem in the decimal form in which it is entered, and write the answer in both the decimal form and converted back to feet and inches.

_____

_____

Unit 30 Area and Pressure Measurement

BASIC PRINCIPLES OF AREA MEASUREMENT

Area is a measure of the size of a surface. It is a two-dimensional measurement. It is expressed in square units such as square inches (in^2), square meters (m^2), or square feet (ft^2). Formulas for the areas of various shapes are given in the chart below. In some instances, it may be necessary to divide a space into smaller pieces, calculate individual areas, and add them to find the total area.

AREA FORMULAS

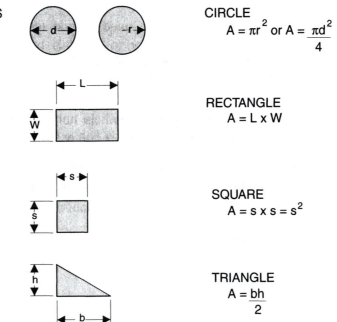

CIRCLE
$$A = \pi r^2 \text{ or } A = \frac{\pi d^2}{4}$$

RECTANGLE
$$A = L \times W$$

SQUARE
$$A = s \times s = s^2$$

TRIANGLE
$$A = \frac{bh}{2}$$

When converting square units to another form, use the same conversion factors as for length measurement, using each one twice. Charts are also available listing some common equivalents; however, they may not include all needed conversions.

Example   Calculate the area of a square 18" on a side in square feet.

$$18" \times 18" = 324 \text{ inches}^2$$

$$\frac{324 \text{ inches}^2}{1} \times \frac{1'}{12"} \times \frac{1'}{12"} = \frac{324}{144} ft^2 = 2.25 ft^2$$

The units of inches could have been converted to feet prior to performing the calculations.

BASIC PRINCIPLES OF PRESSURE MEASUREMENT

Pressure is defined as force applied over an area. It is measured in units of force per unit area. Common units of pressure include pounds per square inch (psi) or pounds per square foot in the U.S. Customary system or Newtons per square meter (pascals) in the SI (metric) system. Stress is an internal reaction of a material to an applied force and is measured in the same type units.

The basic SI unit of force is the newton (N). The U.S. Customary unit of force is the pound. (1 pound-force = 4.448 N.)

PRACTICAL PROBLEMS

For any problems including the value pi (π), use the value 3.14 unless instructed otherwise by your teacher.

1. Find the area of a circle of diameter 3.85 meters. _____

2. Find the area of a rectangle which is 4.75" long and 2.8" wide. _____

3. Find the area of a square with a side length of 24 feet. _____

4. Find the area of a triangle with a base length of 6¼" and a height of 2¾". _____

5. Find the area of a circle with a radius of 5½". _____

6. Find the area of a rectangle which is 275 feet long and 188 feet wide. _____

7. Find the area of a circle with a diameter of 0.45 inch. _____

8. Find the pressure exerted by a 75-pound force over an area of 1.25 square feet. _____

9. Find the pressure exerted by a 220-N force over an area of 0.4 m². _____

10. Find the pressure exerted by a 3,000-pound force over an area of 1.65 ft². _____

11. Find the cross-sectional area of ⅜" × ¾" steel bar, expressed in square inches. Write your answer in both fraction and decimal forms. _____

12. What is the cross-sectional area of 1¼" round bar stock, expressed in square inches? Round to the nearest tenth of a square inch. _____

13. One important factor in making a home energy efficient is eliminating cracks around doors and windows. The illustration below shows a door in a residence.

    a. If there is a ⅛" crack around all sides of the door, calculate the area of the crack in square inches. (Hint: Multiply the length of each side by ⅛" to obtain the area of the crack on that side, add the values for the four sides, and subtract the four "overlap" areas at the corners.)

    _____

    b. If, instead of being a crack around the door, the opening were a square hole in the wall with the same area, what would be the length of a side, to the nearest tenth?

    _____

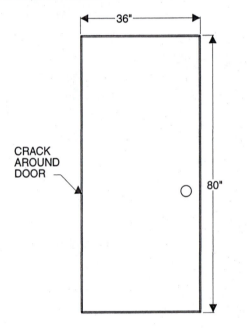

14. A 4" × 4" wood post is used to support a porch roof. If the load carried by the post is 800 pounds, what is the pressure exerted by the post on its support?

    _____

15. Find the cross-sectional area of round tubing with an outside diameter of 1¼" and an inside diameter of 1". (Hint: Find the area of a solid round bar with the same outside diameter and subtract the hollow area.)

    _____

16. The pattern for a cylindrical container, with a diameter of 2" and a height of 3", is shown below.

    a. If the height of the rectangular piece is the height of the cylinder, and its length is equal to the circumference of the circle (C = πd), find its dimensions.   _____

    b. Find the surface area of the cylinder, including the top and bottom.   _____

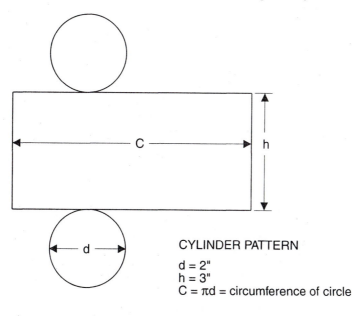

CYLINDER PATTERN

d = 2"
h = 3"
C = πd = circumference of circle

17. What is the square footage of a classroom which measures 38.5' by 28'?   _____

18. A piece of steel bar 1" × ⅜" carried a tensile load of 15,000 pounds. What is the stress in the bar expressed in psi?   _____

19. Convert a pressure measurement of 350 Pa to lb/ft². (1 pascal (Pa) = 1 N/m², and 1 pound-force = 4.448 N.)   _____

20. Nine sacks of sand weighing 80 pounds each are stacked on a wooden cube which is 24" on a side. What is the pressure exerted on the ground by the box in pounds/ft²?   _____

21. For each value shown in the chart below, find the missing equivalent value in the unit specified. Round your answer to the nearest unit.

Convert from:	to:
2 ft^2	in^2
125 in^2	cm^2
9 m^2	yd^2
180 ft^2	yd^2

22. The surface area of a woman's shoe soles is 30 in^2.

    a. If the woman, weighing 135 pounds, distributes her weight equally on the soles, the pressure exerted by the woman on the ground is how many pounds per square inch?   _____

    b. Convert the pressure to pounds per square foot.   _____

23. One way of determining the ductility of a metal is to apply a tensile load to a sample until it fractures. The diameter of the sample is measured before (A_0) and after (A_F) testing. The percent reduction in area is calculated using the expression $\dfrac{A_0 - A_F}{A_0} \times 100$. If $A_0 = 0.610"$ and $A_F = 0.496"$, find the percent reduction in area to the nearest percent.   _____

24. The relationship $P = F \div A$ can be expressed as $F = P \bullet A$ or as $A = F \div P$. A pressure of 50 pounds per square inch (psi) over the face of a hydraulic cylinder piston results in an applied force of 70.6 pounds. Find the area of the piston face and its diameter.   _____

!25. Find the surface area of a closed box which is 5" long, 4" wide, and 3" high.   _____

Unit 31 Volume and Mass Measurement

BASIC PRINCIPLES OF VOLUME MEASUREMENT

Volume is the amount of space occupied by an object and is a three-dimensional measurement. It is measured in units such as cubic inches (in^3), cubic feet (ft^3), and cubic meters (m^3). As an example, the volume of a cube which is 1 foot long, 1 foot wide, and 1 foot high is 1 cubic foot. Formulas which can be used to determine the volumes of various solids are given in the chart below.

VOLUME FORMULAS

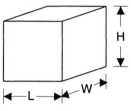

RECTANGULAR PRISM or CUBE

$V = L \times W \times H$

CYLINDER

$V = A \times H$

$A = \pi r^2$

$V = \pi r^2 H$

Conversion factors for several units of volume are listed in a variety of reference sources. They can also be calculated by using a linear conversion factor 3 times, as shown below.

Example: Convert 300 in^3 to ft^3.

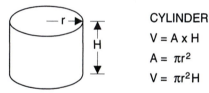

$$300 \text{ inches}^3 \times \frac{1'}{12''} \times \frac{1'}{12''} \times \frac{1'}{12''} = 0.174 \, ft^3$$

Many types of technical problems involve the volume of a fluid. The charts below provide standard volume measures used in the U.S. Customary and Metric systems.

U.S. fluid volume unit	. . . is equivalent to:
1 cup	8 fl oz
1 pint	2 cups
1 quart	2 pints
1 gallon	4 quarts
1 gallon	256 fl oz

U.S. fluid volume unit	Metric equivalent
1 gallon	3.785 liters
1 quart	0.9464 liter

BASIC PRINCIPLES OF MASS MEASUREMENT

The standard unit of mass in the metric system is the kilogram (kg). In the U.S. Customary system, the base unit is the pound, which is a measurement of weight (the effect of the gravitational force on a mass). A constant which is often used to convert between the U.S. and metric systems is *1 pound-mass = 453.6 grams.* This is based on the definition of the mass of the standard pound.

Density is a characteristic of a material which is calculated by dividing mass by volume. The relationship between mass, density, and volume can be expressed in three different ways, as shown below.

$$D = \frac{M}{V}; \quad M = V \times D; \quad V = \frac{M}{D}$$

Example: Find the volume of an aluminum casting if the density of aluminum is 2.7 grams per cubic centimeter (g/cm^3) and its mass is 540 grams.

Using the formula V = M ÷ D, $\quad V = \dfrac{540\ g}{2.7\ \dfrac{g}{cm^3}}$

[Note: Units can be treated as fractions in performing the dimensional analysis calculations. When g is divided by g/cm^3, invert the divisor and cancel factors as needed.] $\quad \dfrac{g}{\dfrac{g}{cm^3}} = g \times \dfrac{cm^3}{g}$

PRACTICAL PROBLEMS

Use 3.14 for π in any problems in which π is used.

1. Find the volume of a cube with sides 3.84" long.   _____

2. Find the volume of a rectangular solid which is 13.6 cm long, 8.5 cm
    wide, and 4.4 cm high. Round to the nearest tenth.   _____

3. Find the volume of a cylinder with a 4.2-foot diameter and a 4.2-foot height.   _____

4. Use the linear relationship of 12" = 1' to find a conversion factor for
    converting in^3 to ft^3.   _____

5. Use the linear relationship of 100 cm = 1 m to find a conversion factor
    for converting cm^3 to m^3.   _____

6. Find the density of an object if its volume is 8.5 cm^3 and its mass is 170 g.   _____

7. Find the volume of an object if its mass is 240 g and its density is 10.5
    g/cm^3.   _____

8. If a 36-cc casting shrinks 4% of its volume as it solidifies, what volume of
    melt must be supplied to the mold cavity to compensate for the shrinkage?   _____

9. If water is flowing over a hydroelectric dam at the rate of 300 ft^3/s, what
    is the flow rate in pounds/s? (1 ft^3 of water weighs 62.4 pounds.)   _____

10. A cylindrical container 4" in diameter must hold a volume (V) of 63 cubic
    inches of liquid.

    a. Find the cross-sectional area (A) of the container using A = πr^2,
        where r is the radius of the circle.   _____

    b. Find the required height (h) of the container using h = V/A.   _____

11. What volume can a pipe with 6" inside diameter, 60' long hold? Express
    the volume in cubic inches and cubic feet.   _____

12. What is the volume of a box 16" long, 14" wide, and 8" high? Express
    the volume in both cubic inches and cubic feet.   _____

13. A cube of aluminum is 24 cm on a side. If the density of aluminum is 2.7
    g/cm^3, what is the mass of the block?   _____

14. A packaging container is 28" long by 20" wide by 12" high. How many packages can be stored in the container if each package is a 4" cube?

    _____

15. How many square feet of concrete slab 3" thick can be poured with 1 cubic yard of concrete?

    _____

16. What is the displacement (D), to the nearest tenth of a cubic inch, for a small gas engine if the cylinder bore is 3.75" in diameter and the stroke length is 3.25"? (The stroke length is the height (h) over which the piston moves. Displacement is calculated using the formula D = Ah, where A is the cross-sectional area of the cylinder.)

    _____

17. The figure below shows two options for an open-top box design which can be cut from like pieces of sheetmetal. Which box will have the larger volume?

    _____

PATTERN DEVELOPMENT FOR OPEN-TOP BOX

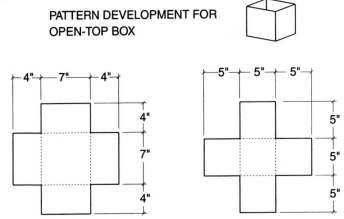

18. A hole is 2' in diameter and 1' deep. A post placed in the hole is 4" x 4". What volume of concrete (in cubic feet) is needed to fill the hole around the post? Round your answer to the nearest cubic foot.

    _____

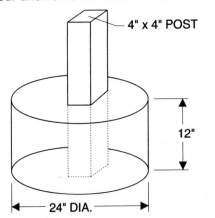

19. A company buys liquid cutting fluid concentrate in 55-gallon drums. Each machine reservoir holds 5 quarts of concentrate. How many machine refills can be made from 3 drums? _____

20. A truckload of sand is 6 cubic yards. How many cubic feet are contained in one load? _____

21. Gasoline is sold in the U.S. by the gallon. It is sold in many other countries by the liter. If a driver purchased 12.6 gallons of gas, what is the equivalent amount in liters? Round to the nearest tenth of a liter. _____

22. An air filtering system is expected to exchange the room air 3 times per hour. If the room is 28' long, 32' wide, and 10' high, how many cubic feet of air must the filter system handle per hour? _____

23. What is the number of 18" × 18" × 18" boxes which can be stacked on a pallet which measures 6' by 6' if the maximum height for stacking is 6'? _____

!24. Review problem 17, which shows a pattern for an open-top box. Complete the chart for boxes of the given dimensions (L, W, and H) which could be built using the full length and width of a 15" × 15" piece of material. Find the volume of each. Then, find the surface area of each box. Which has the largest ratio of volume to surface area? _____

L	W	H	Volume	S. Area	V:SA
13	13	1			
11	11	2			
9	9	3			
7	7	4			
5	5	5			

Unit 32 Energy, Work, and Temperature Measurement

BASIC PRINCIPLES OF ENERGY MEASUREMENT

Energy can be defined as the ability to do work. Energy is measured in joules (1 N-m) in the metric system and in foot-pounds in the U.S. Customary system. A common unit for heat energy is the Btu (1 Btu = 1,055 J).

Example: The kinetic energy of an object in motion is given by the formula $(0.5)mv^2$. If the mass (m) is 8 kg and the velocity (v) is 3 m/s, the kinetic energy is $(0.5)(8 \text{ kg})(3 \text{ m/s})^2 = 36$ kg-m^{2}s^2, which can be "rearranged" into 36 (kg-m/s^2)m = 36 N-m, since 1 N = 1 kg-m/s^2.

BASIC PRINCIPLES OF WORK MEASUREMENT

Work, in the scientific or technical sense, is done when a force is applied over a distance. It can be expressed mathematically as work (W) = force (F) x distance (D). The units for work are joules (1 J = 1 N-m) in the metric system and the foot-pound (ft-lb) in the U.S. Customary system, which are the same units used for energy. Forces are stated in units of newtons (1N=1kg-m/s^2) in the metric system and in pounds (lb) in the U.S. Customary system.

Example: If a 20-pound force is applied to a box, causing it to move 4 feet, the work done is 80 ft-lb.

A related concept is power, which is defined as the rate of doing work. Power is expressed in units of watts (1 W = 1 joule per second or 1 J/s) or horsepower (1 hp = 746 W or $1\text{hp} = \dfrac{1\text{ ft-lb}}{33,000\text{ min}}$).

Notice that when a power unit, with a time unit in the denominator, is multiplied by a time unit, the resulting units are work units.

BASIC PRINCIPLES OF TEMPERATURE MEASUREMENT

Temperature is a measurement of the "hotness" or "coolness" of an object. Several different scales are used to measure temperature, including Fahrenheit, Celsius, and Kelvin.

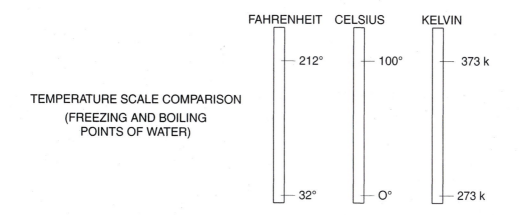

FAHRENHEIT CELSIUS   KELVIN

TEMPERATURE SCALE COMPARISON

(FREEZING AND BOILING
POINTS OF WATER)

212°   100°   373 k

32°   0°   273 k

The relationships between the scales are stated as: $F = \dfrac{9}{5} + 32$; $C = \dfrac{5}{9}(F - 32)$; $K = C + 273$; $C = K - 273$.

Although the Fahrenheit and Celsius scales are more familiar to the general population, all three scales are found in technical and engineering problems. Fahrenheit and Celsius measurements are stated as degrees, but temperature measurements in the Kelvin scale are stated in kelvins, as in 273 kelvins.

Example: Convert 68°F to °C

Since the desired unit is °C, select the formula which has °C isolated: C = 5/9(F-32). Substitute in the temperature in °F and perform the calculations. C = 5/9(68-32) = 5/9(36) = 20°C.

PRACTICAL PROBLEMS

1. Convert 59°F to °C.   _____

2. Convert 80°F to °C. Round to the nearest tenth of a degree.   _____

3. Convert 25°C to °F.   _____

4. Convert 145°C to °F.   _____

5. Convert 15°C to Kelvin.   _____

6. Find the amount of work done in moving a crate 5 feet with an applied force of 45 pounds.   _____

7. Convert 70°F to °C. Round to the nearest degree.   _____

8. If the energy in a 42-gallon barrel of oil is 5.8×10^6 Btu, find the
    equivalent measurement in joules (J).   _____

9. In which situation is more work done, lifting a load 2.5 m using an applied
    force of 35 N, or lifting a load 3.5 m using an applied force of 22 N?   _____

10. The temperature to which a metal must be heated for a hot rolling
    operation is stated as $(0.7)T_M$, where T_M is the melting point of the metal
    on the Celsius scale. For steel with a melting temperature of 1,536°C,
    what is the temperature needed for hot rolling?   _____

11. 1 horsepower is a quantity of power equivalent to 746 watts. If an engine
    develops 3 hp, it is doing work at a rate of how many watts?   _____

12. Engineering calculations often use 5.8×10^6 Btu per barrel of oil and
    27.8×10^3 Btu per kg of coal. How many kg of coal are required to
    produce the same Btu's as 3 barrels of oil? Round to the nearest kg.   _____

13. Car A has a mass of 1,000 kg and is traveling at a velocity of 6 m/s. Car
    B has a mass of 500 kg and is traveling at 12 m/s.

    a. What is the kinetic energy of Car A?   _____

    b. What is the kinetic energy of Car B? (Use the formula $KE = (0.5)mv^2$,
        where KE is kinetic energy, m is mass, and v is the velocity of the
        car.)   _____

14. The melting temperature of aluminum is 660.4°C. Convert the melting
    temperature to °F.   _____

15. The melting point of water is typically stated as 0°C. Write it in °F and in
    Kelvin.   _____

16. A lever is used to lift a 150-lb. motor 3 feet. How much work is done
    lifting the motor?   _____

17. The energy in a 42-gallon barrel of oil is equivalent to 5.8×10^6 Btu. If an
    oil furnace has an efficiency of 65% (converting chemical energy of oil
    into heat), how many Btu of heat are actually output from the furnace per
    barrel of oil?   _____

18. The efficiency of a simple machine is an important characteristic of the machine and is defined as a ratio of the work output to the work input. If a 400-g force is applied over 12 cm to lift a 520-g load over a distance of 5.7 cm, find the efficiency.   _____

19. A pressure of 50 lb/in^2 acts on a piston with an area of 35 in^2.

    a. What is the force applied?   _____

    b. What is the amount of work done when the piston moves 4 inches? Use the formula work (W) = force (F) × distance (D).   _____

20. A block and tackle is used to lift a load of 480 lb over a distance of 1.125 feet by pulling on a rope with a force of 160 lb for a distance of 4 feet.

    a. Find the input work.   _____

    b. Find the output work.   _____

    c. Find the efficiency of the machine.   _____

21. A motor is rated at ⅓ hp. Find its power in watts, to the nearest watt.   _____

Unit 33 Measurements Involving Time

BASIC PRINCIPLES OF MEASUREMENTS INVOLVING TIME

Time is an important component in many technical and engineering problems. Time is involved in a variety of measurements, including velocity, acceleration, power, and RPM. The units of hours, minutes, and seconds are the same in both measurement systems.

Velocity is defined as the change in distance divided by the change in time, or mathematically as $V = \dfrac{d_2 - d_1}{t_2 - t_1} = \dfrac{\Delta d}{\Delta t}$. Note that the *delta* notation (Δ) means "change in" a value.

Example: If a car passes the 288-mile marker at 1:00 p.m. and passes the 312-mile marker at 1:30 p.m., what is its velocity?

$$V = \frac{312 - 288\,\text{mi}}{30\,\text{min}} = \frac{24\,\text{mi}}{30\,\text{min}} = 0.8\,\text{mi}/\text{min}.$$

Acceleration is defined as the change in velocity divided by the change in time, or mathematically as $A = \dfrac{V_2 - V_1}{t_2 - t_1} = \dfrac{\Delta V}{\Delta t}$.

Example: If the velocity of an object increases from 12 ft/s to 30 ft/s in 7.2 seconds, what is its acceleration in ft/s^2?

$$A = \frac{30\,\text{ft}/\text{s} - 12\,\text{ft}/\text{s}}{7.2\,\text{s}} = \frac{18\,\text{ft}/\text{s}}{7.2\,\text{s}} = 2.5\,\text{ft}/\text{s}^2$$

Power is the rate of doing work and is stated in units such as watts (W = J/s) or horsepower. The abbreviation RPM means "revolutions per minute" and is a measure of rotary motion. Another unit for rotary motion is radians per second.

Example: If a 60-watt lamp operates an average of 8 hours per day for 365 days per year, how many watt-hours of energy are used?

Convert the answer to kilowatt hours and calculate the cost to run the light if electrical power costs 10.3 ¢/kWh.

60 W × 8 hr/day × 365 days = 175,200 W-hr = 175.2 kW-hr

175.2 kW-hr × $0.103/kW-hr = $18.05

PRACTICAL PROBLEMS

For problems 1–5, solve for the variable.

1. $V = \dfrac{465 \text{ mi}}{11 \text{h}}$ _____

2. $V = \dfrac{1,760 \text{ ft} - 520 \text{ ft}}{80 \text{ s}}$ _____

3. $A = \dfrac{84 \text{ ft/s} - 50 \text{ ft/s}}{10.5 \text{s} - 3.3 \text{ s}}$ _____

4. $R = 1,150 \text{ rpm} \times 24 \text{ min.}$ _____

5. $A = \dfrac{83 \text{ ft/s} - 27 \text{ ft/s}}{27 \text{s} - 13 \text{ s}}$ _____

6. Find the velocity if a car travels 208 miles in 3.5 hours. _____

7. Find the velocity if an assembly line moves 65 feet in 8.3 seconds. _____

8. Find the acceleration if the velocity of an assembly line conveyor belt increases from 12 ft/s to 18 ft/s in 4 seconds. _____

9. A 100-watt light bulb uses electrical energy at a rate of 100 J/s. If the bulb operates for 2 hours, how many joules of energy would be used? _____

10. A large air cushion vehicle can travel at 65 knots. Find the equivalent speed in miles per hour and kilometers per hour. 1 knot = 6,080 ft/h. _____

11. A factory has a just-in-time delivery agreement with its supplier.

    a. If the supplier is to have a truck at the factory at 3 p.m. on Monday, the distance to be traveled is 165 miles, and the truck averages 55 miles per hour, at what time should the truck leave the supplier? _____

    b. If the supplier wants to allow 45 minutes in case of heavy traffic or other problems, when should it leave? _____

12. An animation program displays 24 frames per second. How many frames would be displayed in a 1-minute presentation? _____

13. If a machine is running at 225 rpm, how long will it have to run to make 2,800 revolutions for a test?   _____

14. How many joules of energy are delivered in 1 hour by an engine developing 3 hp? Express your answer in scientific notation.   _____

15. A car reaches a velocity of 65 miles per hour in 6.2 seconds, starting at 0 mph. What is the acceleration of the car in ft/s²?   _____

16. A car travels 924 feet in 22 seconds. What is the velocity?   _____

17. An elevator lifts a load weighing 80 N vertically 140 m.

   a. How much work is done?   _____

   b. If the time required is 25 seconds, what is the power, expressed in horsepower?   _____

18. Express an angular (rotary) speed of 250 rpm as radians/second.   _____

19. Find the velocity of a conveyor belt which moves 46 feet in 16 seconds.   _____

20. When the brake is activated, the velocity of a car decreases from 65 mph (miles per hour) to 30 mph in 5.5 seconds. Find the deceleration in ft/s². Be sure to convert units as needed.   _____

21. Express an angular (rotary) speed of 50 radians per second in rpm.   _____

!22. Describe the error that was made in the conversion from miles per hour to feet per second.

$$\frac{30\text{ mi}}{1\text{ hr}} \times \frac{60\text{ min}}{1\text{ hr}} \times \frac{60\text{ s}}{1\text{ min}} \times \frac{5{,}280\text{ ft}}{1\text{ mi}}$$

Unit 34 Review and Combined Problems on Measurement

PRACTICAL PROBLEMS

For problems 1–10, convert each measurement to its equivalent. Round to 3 decimal places if needed.

1. 105 in. = _____ ft.   _____

2. 7 ft. 3 in.= _____ m   _____

3. 14 in^2 = _____ cm^2 _____

4. 6 yd^3 = _____ m^3 _____

5. 55 mi/h = _____ km/h   _____

6. 55 mi/h = _____ ft/s   _____

7. 32 gallons = _____ liters   _____

8. 6.4 yd^2 = _____ ft^2 _____

9. 105 degrees = _____ radians   _____

10. π/6 radians = _____ degrees   _____

11. 45 pounds/ft^2 = _____ N/m^2 _____

12. Find the area of a circle of diameter 2.15 meters.   _____

13. A bill of materials for a cabinet is shown. Complete the bill of materials by calculating the cost for each item and the total cost.

BILL OF MATERIALS FOR CABINET

QTY.	DESCRIPTION	UNIT COST	COST
2	¾" Oak lbr. core plywood	44.75	
2	1"x6"x8' oak lumber	12.85	
4 pr.	⅜" offset brass hinges	3.80/pr.	
1	3' X 8' almond laminate	2.05/sq ft	
3	1" round brass knob	1.50	
1	20" drawer glide set	4.65	
		TOTAL	

14. A computer network wiring job requires 2 pieces of plenum cable 85' long, 8 pieces of plenum cable 28' long, 10 pieces of thin Ethernet cable 8' long, 1 piece of plenum cable 40' long, and 2 pieces of thin Ethernet cable 6' long.

a. How much plenum cable is needed for the installation? _____

b. How much thin Ethernet cable is needed? _____

15. a. What is the cross-sectional area of a 2×4 if the actual dimensions are 1½" by 3½"? _____

b. If the board originally was actually 2" by 4" and was reduced by finishing operations on all four sides, use the original cross-sectional area and the finished cross-sectional area to find the amount of waste as a percentage of the original area. _____

16. If a 9" by 12" column is supporting a load of 4,200 pounds, what is the pressure (in pounds/in^2) exerted by the column on its base? Express your answer to the nearest tenth. _____

17. What is the area of a building which measures 42' by 68'? _____

18. Convert a pressure measurement of 1 pascal (Pa) to lb/in^2. (1 Pa = 1 N/m^2). _____

19. What is the volume, to the nearest cubic inch, of a cylindrical container 6" in diameter and 8" high? _____

20. How many square feet of concrete slab 4" thick can be poured with 1 cubic yard of concrete? _____

21. A cube of copper is 6 cm on a side. If the density of copper is 8.93 g/cm^3, what is the mass of the block? _____

22. Find the volume of an object if its mass is 180 g and its density is 11.3 g/cm^3. _____

23. Primer is used at the rate of 3.5 quarts per day in a finishing department. How many full days' supply is one 55-gallon drum? _____

24. A gas tank on a mid-sized car holds 16 gallons. What is the tank capacity in liters? _____

25. How many joules of energy are delivered in 1 hour by an engine developing ⅓ horsepower? _____

26. A car reaches a velocity of 40 miles per hour in 3.8 seconds, starting at 0 mph. What is the average acceleration of the car in ft/s^2? Round to the nearest tenth. _____

27. A second-class lever is used to lift a load of 85 pounds a height of 9 inches. How much work is done (in ft-lb)? _____

28. The temperature of a chemical tank in a plating department must be maintained at 185°F. What is the temperature in °C? _____

29. An air filtering system must exchange the room air 3 times per hour. If the room is 12' wide, 14' long and 8' high, how many cubic feet of air must the filter system handle per hour? _____

30. Convert 35°C to °F. _____

31. A belt on a sander makes 10 revolutions in 7.5 seconds.

    a. What is its angular speed in rpm? _____

    b. If the circumference of the belt is 42", what is the speed in feet per minute? _____

32. The mechanical work (W) done by a pump is the product of the weight (W) of the fluid moved and the height (h) to which it is lifted. If a pump moved 5 gallons of water to a container which was 3 feet higher than the water source, how much work did the pump do? (Hint: Convert gallons to pounds of water; a gallon of water is 0.134 ft^3 and water weighs 62.4 lb/ft^3.)

          _____

33. A tanker truck carrying 2,280 gallons was loaded in 25.5 minutes. What was the flow rate in gallons per minute?

          _____

Tables, Charts, and Graphic Representation of Data

Data and information in technical and engineering problems are often shown using tables, charts and graphs. Tables are found in many sources, including handbooks and other reference materials.

There are three main forms of graphic representation used in business and industry: pie graphs, line graphs, and bar graphs. The type of graph or chart selected is based on the type of information to be presented. Pie graphs are best suited for percentage data, line graphs are effective in comparing values in two variables such as sales over time, and bar graphs are good for comparing relative values. Some data can be presented effectively on more than one type of graph. An example of each type is illustrated below.

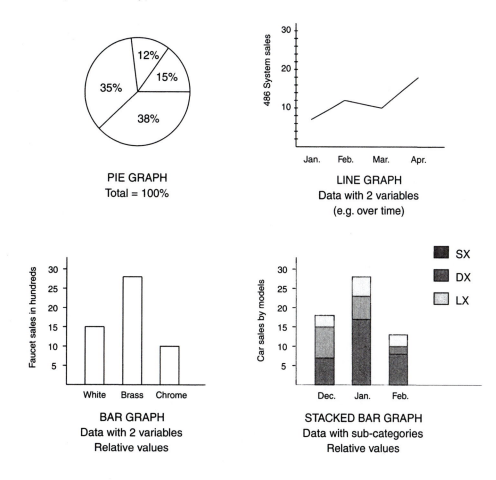

PIE GRAPH
Total = 100%

LINE GRAPH
Data with 2 variables
(e.g. over time)

BAR GRAPH
Data with 2 variables
Relative values

STACKED BAR GRAPH
Data with sub-categories
Relative values

Unit 35 Line Graphs

BASIC PRINCIPLES OF LINE GRAPHS

Data can be presented using line graphs, which is similar to graphing ordered pairs on a Cartesian coordinate system. The Cartesian system uses two perpendicular axes that divide a plane into four quadrants. The first quadrant is used when displaying data with positive values or when using a CAD system for two-dimensional drawings.

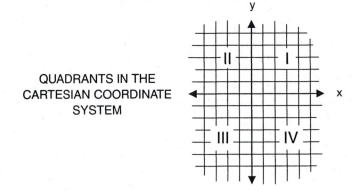

QUADRANTS IN THE
CARTESIAN COORDINATE
SYSTEM

Graphing involves identifying two variables which are characteristic of the data items and representing a data point by its values for those two variables. Graphs should be drawn to an appropriate scale to illustrate the relationships between the data values, although the scale may be different on each axis. In the Cartesian coordinate system, the horizontal axis is called the *x-axis,* while the vertical axis is called the *y-axis.* The intersection of the two axes is called the *origin.* Each point is labeled with an *ordered pair,* with the origin at (0,0). Note that the x-value is always listed first in an ordered pair.

Example: Locate the points A (3,5), B (6,2), and C (4,4) on the coordinate system.

The point (3,5) is located by starting at the origin and moving 3 units to the right (x-value) and 5 units up (y-value). Locate the other points using the same process.

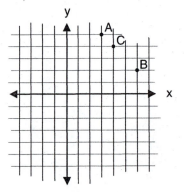

PRACTICAL PROBLEMS

1. Locate the points (2,6), (4,1) and (5,4) on the coordinate system.

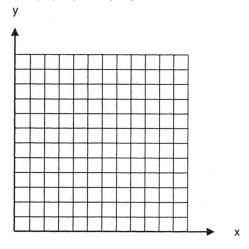

2. Write the coordinates of points A, B, C, and D as shown on the coordinate system.

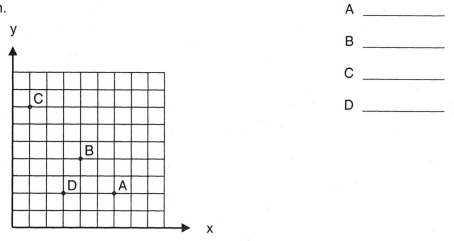

A _____

B _____

C _____

D _____

3. CAD systems use two types of coordinates: absolute and relative. Absolute coordinates are used when a point is located with respect to the origin. The ordered pair representing a point is used, similar to graphing in math class. For the drawing below, write the ordered pair which represents each labeled point.

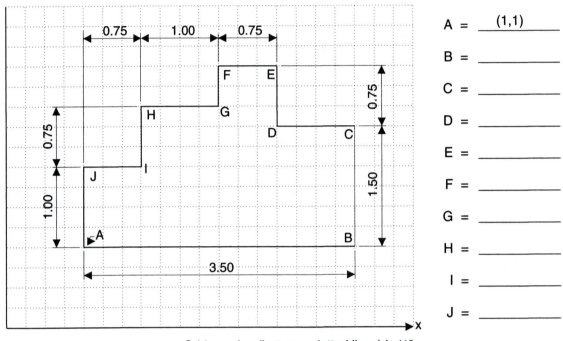

A = ___(1,1)___

B = _____

C = _____

D = _____

E = _____

F = _____

G = _____

H = _____

I = _____

J = _____

Grid spacing (between dotted lines) is ¼"

4. Relative coordinates are used in CAD when the location of a point is described with respect to another point. In other words, to get to point B from point A, a displacement written as R(3,2) means that point B is 3 units to the right and 2 units up from point A. (Different CAD programs use different ways of indicating relative coordinates, but R will be used in this example.) *Use a negative sign to indicate a movement to the left for the x coordinate and down for the y coordinate.* Using the part drawing below, write the coordinates for each point listed relative to the previous point.

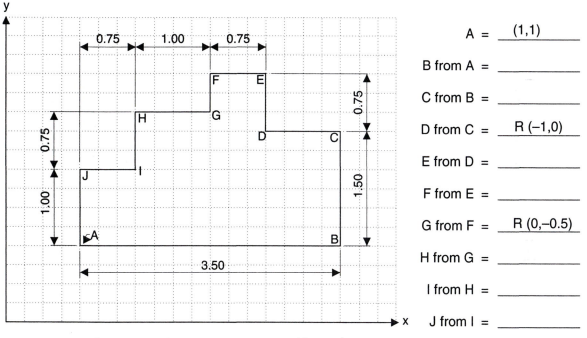

A = ___(1,1)___

B from A = _____

C from B = _____

D from C = ___R (−1,0)___

E from D = _____

F from E = _____

G from F = ___R (0,−0.5)___

H from G = _____

I from H = _____

J from I = _____

Grid spacing (between dotted lines) = ¼"

5. The line graph below shows production times for the first 6 parts produced in a run, with the x-value representing the part number and the y-value representing the dimension. For each point, record the time to the nearest half-hour.

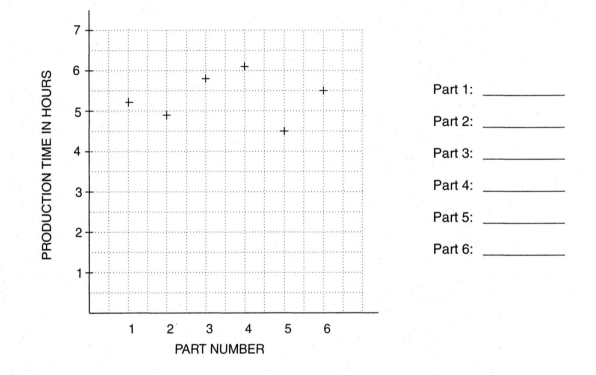

Part 1: _____

Part 2: _____

Part 3: _____

Part 4: _____

Part 5: _____

Part 6: _____

6. The chart below shows the velocity of a moving object versus the time in
    seconds. Using the chart, estimate the speed of the object at 7 and 10
    seconds.

    7s _____

    10s _____

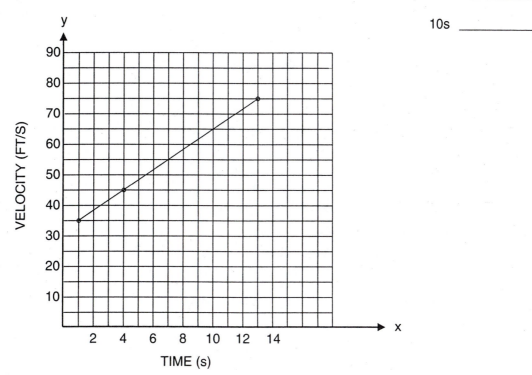

7. On the given coordinate system, title the x-axis "Departments." The y-axis title should be "Days Without a Lost-Time Accident." Determine an appropriate scale for the vertical axis and label the scale. Label the department titles along the x-axis. Then plot the following data points on the graph.

Plating 75	Machine Shop 54
Painting 145	Welding 180

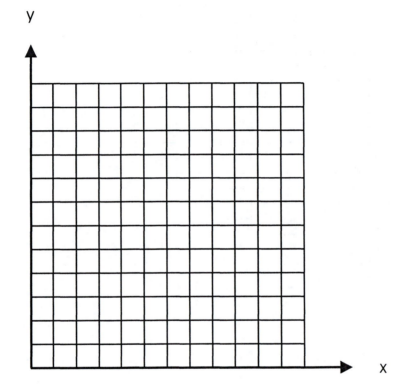

!8. A chart known as the X-bar chart is often used in quality control. The measured dimension for each part is plotted in sequential order on a chart, with the x-value representing the part number and the y-value representing the dimension. For each part dimension listed, plot the value on the graph shown. Then calculate the average of the dimension values to the nearest thousandth and draw a dotted line across the graph representing that value. Values which are above the upper limit or below the lower limit are out of specification. Which parts were out of specification?

_____

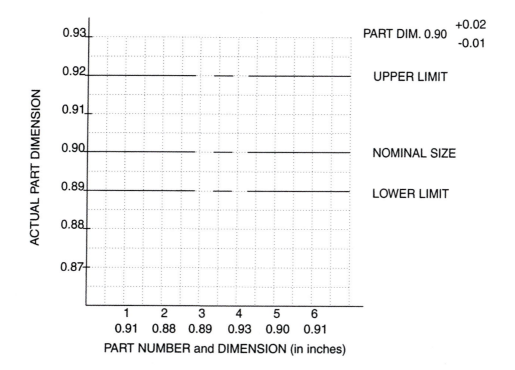

PART DIM. 0.90 +0.02
                -0.01

UPPER LIMIT

NOMINAL SIZE

LOWER LIMIT

ACTUAL PART DIMENSION

1	2	3	4	5	6
0.91	0.88	0.89	0.93	0.90	0.91

PART NUMBER and DIMENSION (in inches)

Unit 36 Pie Graphs

BASIC PRINCIPLES OF PIE GRAPHS

Pie graphs are typically used when data is in percentage form. The entire circle on a pie graph represents 100%. The graph is divided into wedges, each representing a percentage of the total. Since 100% equals the total of the data, and a circle has 360°, each one percent is represented by 3.6°.

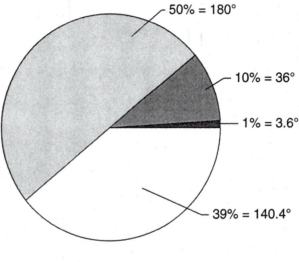

PIE GRAPH

A protractor or circle graph paper can be used to locate the lines which divide the circle into parts.

Example: In a vocational math class, a survey was made to identify the major field of study of each student. The results were: carpentry (8), diesel mechanics (8), welding (6), drafting (12), and HVAC (6). Draw a pie graph showing the fields of study.

PIE GRAPH OF CLASS
MEMBERS BY MAJOR
FIELD OF STUDY

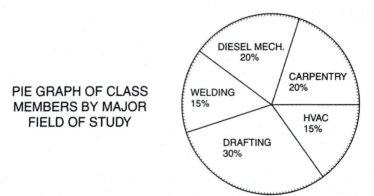

The total number of students in the class is 40. The angle of each piece of the graph can be calculated by multiplying the percentage (in decimal form) by 360 degrees. For carpentry and diesel mechanics, $0.20 \times 360° = 72°$ each. For welding, $0.15 \times 360° = 54°$. The other segments are calculated using the same process.

CALCULATOR USE

Some calculators have a percentage key [%]. When a number is entered and the [%] key is pressed, the number is converted into a percentage which can be used in calculations. As an example, 36 can be increased by 25% by entering the sequence 36 [+] 25 [%] [=]. The display should read 45. Operations other than addition can also be done using similar procedures. Consult the owner's manual for your calculator.

PRACTICAL PROBLEMS

1. Convert each percentage to decimal form.

   27% _____

   45% _____

   8% _____

   16% _____

   4% _____

2. a. What is 35% of 360°? _____

   b. What is 88% of 360°? _____

3. An office building had a total of 35 printers on a network, including 8 laser printers and 11 inkjet printers.

   a. If the rest of the printers were dot-matrix, find that quantity. _____

   b. Calculate the percentage of total printers for each printer type.   Laser _____

   Inkjet _____

   Dot-matrix _____

4. For the data below, find the percentage of the total quantity and pie chart degrees for each product. Then match each product to the appropriate section of the pie graph. Write the letter of the correct section in the chart.

PRODUCT	QTY	LTR	%	DEG
2-HANDLE KITCHEN W/ SPRAY	42			
2-HANDLE KITCHEN W/O SPRAY	84			
SINGLE HANDLE KITCHEN W/SPRAY	118			
SINGLE HANDLE KITCHEN W/O SPRAY	101			
2-HANDLE W/ HIGH-RISE SPOUT	29			
SINGLE HANDLE W/HIGH-RISE SPOUT	42			

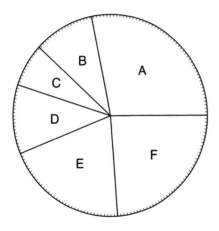

Each division equals 2°.

5. For each data entry and percentage in the chart below, calculate the number of degrees in a pie chart which would be needed to represent the percentage, to the nearest degree. Check your calculations using the Total row.

Product	Percent of sales	Degrees in chart
20-amp breaker	22%	
15-amp breaker	48%	
duplex outlet	20%	
3-way switch	10%	
Total	100%	

6. For the data below, find the percentage of total sales of each program (spreadsheet, etc.) and calculate the angle appropriate for each one on a pie chart. Represent the data in a pie chart.

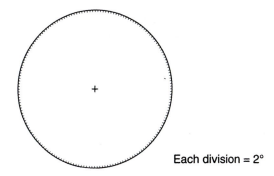

Each division = 2°

Program	3.5" disk	5.25" disk	CD-ROM	Program Sales	% of Sales	Degrees on chart
Spreadsheet	200	60	100			
Word processor	600	180	240			
Database	150	140	130			
Total						

7. Sales data for a year, grouped by product line, is given in the chart below. Represent the data in a pie graph.

Product	Qtr1	Qtr2	Qtr3	Qtr4	Total	% of Sales
Single deck cassette	400	310	280	360		
Double deck cassette	850	760	520	550		
CD with cassette	510	480	670	750		
				Total		

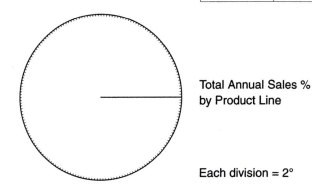

Total Annual Sales % by Product Line

Each division = 2°

8. A machine shop recorded data, classifying rejected parts according to the type of problem which caused the part to be rejected. The various categories and number of parts rejected for each problem were: holes drilled off-center (15), over-sized drilled holes (12), undersize diameter on turned parts (27), wrong thread (6), and improper chamfer (10). Draw a pie graph representing the data.

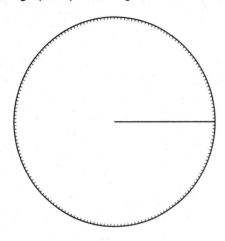

Each division equals 2°.

Unit 37 Bar and Stacked Bar Graphs

BASIC PRINCIPLES OF BAR GRAPHS

Bar graphs are appropriate when comparing quantities. They do not require that the quantities add up to a total value or 100%.

Stacked bar graphs are a variation of bar graphs in which each bar represents a total value, with each bar divided into sub-parts. An example of each is shown below.

A bar graph uses the height or length of bars to compare the values of data. In order to accurately represent the relationship between values, the numerical axis should begin at zero, although this may not be practical for large values. (Note: Many bar graphs used in printed media do not start at zero, which can distort the relationship between values visually. It is important to look carefully at the scale of the numerical axis.)

PRACTICAL PROBLEMS

1. For the bar graph shown, fill in the data to the nearest 100 in the spaces provided.

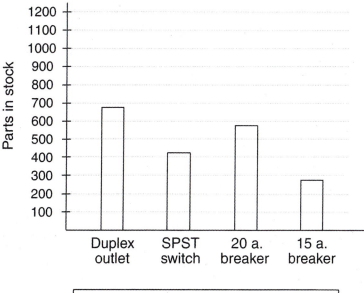

Electrical Parts Inventory			
D.Out.	SPST	20a.br.	15a.br.

2. The overhead costs for a manufacturing plant are listed by year. On the graph form below, draw a bar representing each year. Shade each bar according to the key provided.

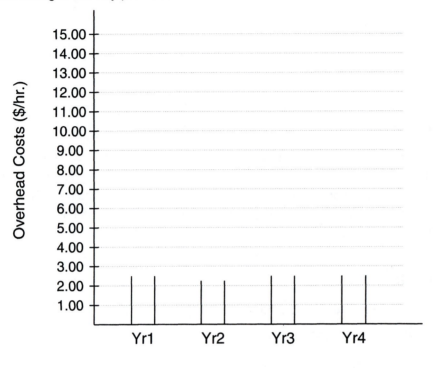

Hourly Overhead Costs			
Year 1	Year 2	Year 3	Year 4
$11.75	$13.00	$12.50	$14.00

3. Using the data in problem 6 in Unit 36, draw a stacked bar chart for each form of media, with the parts of each bar representing the program. Use the key provided.

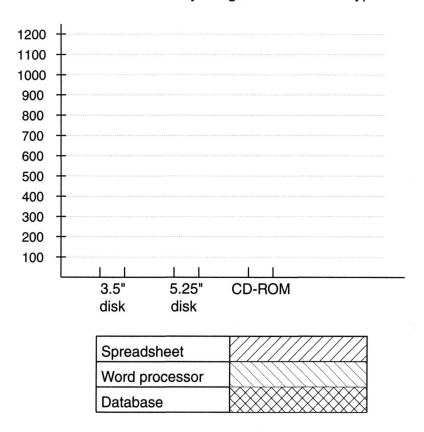

Software Sales by Program and Media Type

4. For the data and bar chart provided, write in the correct labels for the bars.

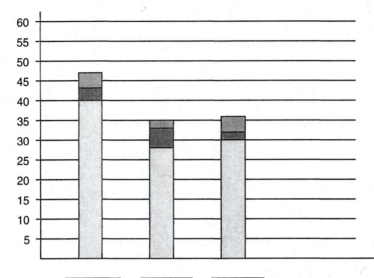

Accepted	
Reworked	
Rejected	

Part No.	Accept	Rework	Reject
27-162	28	2	5
27-155	40	4	3
29-351	30	4	2

5. The bar graph below shows the number of products of each model shipped during a 6-month period.

   a. Visually, how does the height of bar 2 compare with the height of bar 3?

   _____

   b. What part of the graph shows that the actual ratio of bar 2 to bar 3 is not the same as the visual appearance?

   _____

   c. What are the actual values of bars 2 and 3 and the ratio between the values?

   _____

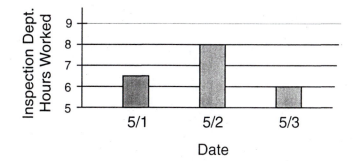

Formulas and Equations

Formulas and equations are very common in engineering and industrial technology applications. There are many types of equations, including first degree, second degree (quadratic), and equations of higher degree. First degree equations contain variables raised only to the first power.

Examples: $k + 37 = 75$

$d = r \bullet t$

The equation $y = x^2$ is not a first degree equation, since x is raised to the second power (has an exponent of 2).

In this unit, first degree equations will be written and solved. A variety of equations and formulas will be presented, including some in general use and some which are more specific to Industrial Technology.

Unit 38 Representation in Formulas and Equations

BASIC PRINCIPLES OF FORMULAS AND EQUATIONS

A formula is a math statement or equation expressing the relationship between related values. Several formulas, including those for temperature conversion, kinetic energy, and work have been introduced in problems in earlier units. Standard formulas are typically obtained from reference materials such as the appendices of textbooks, trade handbooks, or technical manuals published by trade or professional associations.

Formulas contain variables, constants, and math operation symbols. Constants are numbers or values that do not change, like π (the ratio between the circumference and diameter of a circle) and g (the force of gravity on an object). Variables represent values that change under different situations, like v (velocity) and R (resistance). Standard symbols for variables have been identified for many types of data and are used and recognized by students, engineers and technologists.

REPRESENTING PRACTICAL PROBLEMS IN MATHEMATICAL FORM

In practical problems, a verbal description of a problem must be translated into variables and math symbols. Many of the terms used in verbal problems and their symbol representation were introduced in earlier units and should be reviewed as needed. In this unit, the emphasis will be on translating problems into math equations and expressions.

When expressing a practical problem as an equation, it is important to *express each unknown as a variable,* and *label the quantity that it represents.* The variable and any constants are combined with the appropriate math symbols. It may also be helpful to draw a diagram to help describe the relationships in the problem.

 CALCULATOR USE

Calculators can be used in finding the solutions to equations. One method is to write the equation so that the variable is isolated on one side of the equation, and enter the math expression on the other side of the equation into the calculator, being careful to follow the Order of Operations. It is best if the entire expression can be entered as one series of keystrokes to avoid rounding errors.

PRACTICAL PROBLEMS

For problems 1–11, write a mathematical formula or equation. Use the symbols stated in the problem.

1. The power (P) used by a circuit is equal to the voltage (E) squared divided by the resistance (R). _____

2. The kinetic energy (KE) of an object in motion is equal to one-half of the mass (m) multiplied by the square of the velocity (v). _____

3. The energy efficiency (e) of an energy conversion process is a ratio of the output energy (E_o) to the input energy (E_i). _____

4. The total cost (C) of a repair job is the sum of the material cost (M) and the total labor cost (L). The total labor cost (L) is the product of the number of hours worked (n) and the hourly rate (r). (Write two equations and then substitute the equivalent expression for L into the equation for total cost.) _____

5. The radius (r) of a circle is one-half the diameter (d). _____

6. A temperature measurement in Fahrenheit degrees (°F) is equal to ⁹⁄₅ multiplied by the Celsius reading (°C), plus 32 degrees. _____

7. The Pythagorean theorem states that the square of the length of the hypotenuse (c) of a right triangle is equal to the sum of the squares of the lengths of the other two sides (a and b). _____

8. For a lever, the effort force (E) multipled by its distance from the fulcrum (ED) is equal to the resistance force (R) multiplied by its distance from the fulcrum (RD). _____

9. For a circuit containing 3 components connected in series, the total resistance (R_T) in the circuit is equal to the sum of the individual resistances (R_1, R_2, and R_3). _____

10. The area (a) of a piece of sheetmetal is the product of the length (L) and width (W). The cost (c) of a piece of sheetmetal is the product of the area (a) and the price per square foot (p). Write a formula for the cost of a piece of sheetmetal if the price per square foot is $0.88. _____

11. In tensile testing of materials, the percent elongation (e) is 100 multiplied by a ratio of the deformation length (Δl) of the specimen to the length before testing (l_i). The deformation length (Δl) is the difference of the length after testing (l_f) and the length before testing (l_i). (Write two equations and substitute the equivalent expression for deformation length into the equation for percent elongation.)

_____

_____

For problems 12–20, in the space below each math equation, write a verbal "equation" for each math equation or formula.

12. $E = I \cdot R$, where E = voltage, I = amperage, and R = resistance.

13. $I = E \div R$, where E = voltage, I = amperage, and R = resistance.

14. $C = 512n$, where C = capacity of a floppy disk in bytes, n = number of sectors on the disk, and 512 is the constant number of bytes per sector.

15. $KE = \frac{1}{2}mv^2$, where KE = kinetic energy, m = mass of the moving object, and v = velocity of the object.

16. $P = I \cdot E$, with P = power, I = amperage, and E = voltage in an electrical circuit.

17. $S = hr$, where S = gross salary, h = number of hours worked, and r = hourly rate of pay.

18. $a = (v_f - v_i) \div t$, where a = acceleration, v_f = final velocity, v_i = initial velocity, and t = time.

19. $C = \frac{5}{9}(F - 32)$, where C = temperature in $°C$ and F = temperature in $°F$.

20. $A = \pi r^2$, where A = area of the circle and r = radius of the circle.

Unit 39 Solving Equations

BASIC PRINCIPLES OF EQUATIONS

An equation is a statement that two math expressions are equal in value. An equation may be thought of as a "balanced" sentence, with the equal sign serving as the fulcrum or balance point for the scale.

$$\frac{\text{expression} = \text{expression}}{\Delta}$$

Using that image, remember that any math operations that are performed on one side of the equation must also be performed on the other side. The equation must remain "balanced" or equal in value on each side, although its appearance may change.

VARIABLES VS. CONSTANTS

A constant is a symbol which always has the same meaning, such as 3, 475, or π. A variable is a symbol, typically a letter or Greek letter, which represents an unknown quantity. If a formula is to be solved to find the numerical value of a variable, there must be only one variable. If the formula contains more than one variable, it is sometimes possible to solve for the value of a variable in terms of the other variables. That process is often called solving a *literal equation.*

STRATEGIES FOR SOLVING AN EQUATION

One of the main strategies for solving an equation is to isolate the variable on one side of the equation and all constants and/or other variables on the other side. In order to do that, keep in mind that the order of operations must be followed. (If needed, review the order of operations in Unit 5.)

When isolating a variable, the math operations which affect the variable must be "undone" in the reverse order in which they would be applied according to the order of operations. Each math operation must be "undone" using its inverse operation. For example, the inverse operation of addition is subtraction. The inverse operation for multiplication is division, and for a power is the appropriate root.

Example: Solve $E = I \times R$, for I

$$E = I \times R \qquad \frac{E}{R} = \frac{I \times R}{R} \qquad \frac{E}{R} = I$$

Example: Solve $A = \pi r^2$, for r

$$A = \pi r^2$$
$$\frac{A}{\pi} = \frac{\pi r^2}{\pi}$$
$$\sqrt{\frac{A}{\pi}} = \sqrt{r^2}$$
$$\sqrt{\frac{A}{\pi}} = r$$

Example: Solve $2(3x + 7) = 38$, for x

$$2(3x + 7) = 38$$
$$6x + 14 = 38$$
$$6x + 14 - 14 = 38 - 14 = 24$$
$$x = 24 \div 6 = 4$$

When a practical problem includes units, be sure to include the units when you write the math equation, and determine the correct units using the procedure presented in Section 8.

Note: If a problem contains fractions, the fractions can be eliminated by finding a common denominator for *all* the fractions and multiplying both sides of the equation by that number.

PRACTICAL PROBLEMS

For problems 1–8, solve the literal equation for the specified variable.

1. Solve $E = I \times R$, for R. _____

2. Solve $C = 512n$, for n. _____

3. Solve $P = I \times E$, for E. _____

4. Solve $S = hr$, for h. _____

5. Solve $a = \dfrac{(V_f - V_i)}{t}$, for v_f. _____

6. Solve $KE = \frac{1}{2}mv^2$, for v. _____

7. Solve $C = \dfrac{5}{9}(F - 32)$, for F. _____

8. Solve $a = \dfrac{(V_f - V_i)}{t}$, for t.

_____

9. The power used by a heating element is 2,400 watts. If the voltage is 220 volts, what is the resistance of the element? Use the formula $P = E^2/R$.

_____

10. What is the mass of an object in motion if its velocity is 3 m/s and its kinetic energy is 180 J? Use the formula $KE = (0.5)mv^2$.

_____

11. An electrician worked at a repair job for 6.5 hours and earns $11.50 per hour. If the cost of materials for the job was $48.25, what is the total cost of the repair?

_____

12. A cabinet installation job required a piece of oak veneer which was 24" by 42". If the veneer is priced at $4.85 per square foot, what is the cost of the veneer?

_____

13. Room temperature is often assumed to be 70° F. What is the equivalent temperature expressed to the nearest degree Celsius? Use $C = \frac{5}{9}(F-32)$.

_____

14. Using a first-class lever, an effort force (E) of 22 pounds is placed 7.5 feet from the fulcrum (ED) and used to lift a load of 105 pounds (R). To the nearest tenth of a foot, at what distance from the fulcrum (RD) was the resistance or load? Use the relationship "Effort force multiplied by effort distance is equal to resistance force (load) multiplied by resistance distance."

_____

22 lb

105 lb

7.5 ft

15. A circuit contains 4 components in series with resistances of 3.5 ohms, 7.25 ohms, 12 ohms, and 6.4 ohms. What is the resistance of the circuit if the total resistance of components in series is the sum of the individual resistances?

_____

16. A test specimen of a metal is tested for ductility using tensile testing. If the original length (l_0) was 2.5 inches and the length after testing (l_F) was 2.73 inches, what is the percent elongation? Use % elongation = $(l_F\text{-}l_0)/l_0 \times 100$.

_____

17. The relationship between number of teeth (T or t) and number of turns (N or n) for two meshed gears is stated as TN=tn, with capital letters used for one gear and lower-case for the other. If one gear has 52 teeth and turns at 120 rpm, and the other gear turns at 480 rpm, how many teeth are on the second gear?

_____

18. In one text, the formula for circular area is given as A = $(0.7854)d^2$, while in another text it is given as A = πr^2. Use equation solving techniques, 3.1416 for π, and the relationship between radius and diameter to show that they are basically the same.

_____

19. Using the formula in problem 7, convert the melting temperatures of the metals listed in the table to the missing form. Round to the nearest tenth.

Metal	Melting temp (°F)	Melting temp (°C)
Aluminum	1220.7	
Copper		1084.9
Gold	1947.9	
Lead		327.4

20. A vehicle's acceleration is tested by measuring the time it takes to change from one velocity to another. If the initial velocity (v_i) is 20 mph and the final velocity (v_f) is 78 miles per hour, with the time (t) of 4.3 seconds required, what is the acceleration (a) in ft/s^2? Use the formula $a = \dfrac{v_f - v_i}{t}$ and round to the nearest foot/s^2.

_____

21. Angular measurement can be made in two systems, degrees and radians. The relationship between the systems is 2π radians = 360°. Rewrite the formula to solve for 1 radian in terms of degrees.

_____

22. If the diameter of a drilled hole is 0.875 inch, what is its radius to the nearest thousandth?

_____

23. For a circuit with 3 resistors in parallel, the formula for total resistance is $\frac{1}{R_T} = \frac{1}{R_1} + \frac{1}{R_2} + \frac{1}{R_3}$. If $R_1 = 20$ ohms, $R_2 = 80$ ohms, and $R_3 = 16$ ohms, find the total resistance.

_____

!24. The cutting speed for a machining process such as turning on a lathe or drilling is expressed in surface feet per minute. The correct rpm setting for the machine is determined using the formula $rpm = \dfrac{CS' \times \frac{12''}{1'}}{D'' \times \pi}$. It is often approximated by the formula $rpm = \dfrac{CS' \times 4}{D''}$. Using equation solving techniques and the value of π, explain how the second formula was derived and why it is useful as a close approximation.

_____

!25. In Unit 19 on Proportion, cross-multiplication was used. Verify that it is mathematically correct by following the steps listed for the equation $\frac{a}{b} = \frac{c}{d}$.

a. Find a common denominator for both fractions.

_____

b. Multiply both sides by that denominator.

_____

c. Divide out common factors.

_____

!26. The formula for compound interest is stated as $A = P (1 + i)^n$, where A is the total amount after compounding, P is the principal, i is the annual interest rate and n is the number of years the money is invested. Find the amount in a savings account if $4,250 was invested for 5 years at 7.75% annual interest.

_____

!27. Using the compound interest formula and amounts in problem 26, adjust the formula to show interest compounded quarterly. (Change i to a quarterly interest rate, to 5 decimal places, and change n to the number of quarters for which the money was invested.)

a. What is the total amount in the account?

_____

b. What is the difference in the two amounts, with annual versus quarterly compounding?

_____

!28. Using the compound interest formula and amounts in problem 26, adjust the formula to show interest compounded daily. (Change i to a daily interest rate, to 5 decimal places, and change n to the number of days for which the money was invested.)

   a. What is the total amount in the account?   _____

   b. What is the difference in the two amounts, with annual versus daily compounding?   _____

Unit 40 Formulas Common in Industrial Technology

PRACTICAL PROBLEMS

Use the formulas in the Appendix as needed.

1. The recommended cutting speed for a finish cut on aluminum with an HSS lathe tool is 300 surface ft/min. For a ¾" diameter, find the correct rpm to the nearest whole number.

   _____

2. If the Pythagorean theorem is being used to square the forms for a building slab, and the measurements down each perpendicular side are 15 and 20 feet, what should the distance between the side measurements (the hypotenuse) be?

   _____

3. A board foot is a lumber measurement. The nominal size of lumber is its "name", such as 2 × 4, although the nominal size may differ from its actual measurement.

   a. How many board feet are in a 1 x 8 × 4'?

   _____

   b. How many board feet in a 1 × 6 × 10'?

   _____

4. A furniture builder ordered oak lumber from a supplier. If she ordered 8 pieces of 1"×6"×8' lumber and 12 pieces of 1"×8"×6' lumber, how many board feet are in the order?

   _____

5. The recommended cutting speed for a heavy cut on low carbon steel with a carbide tool on a lathe is 350 surface feet per minute. To turn a 1.5" diameter part, what is the correct rpm to the nearest hundred?

   _____

6. Resistance in a length of wire is expressed by the formula $R = \dfrac{\rho \times l}{A}$,

   where ρ is the resistivity of the wire material in ohm-cm, l is the length (in cm) and A is the cross-sectional area of the wire (in cm^2). If the resistivity of copper is 1.724×10^{-6}, and the wire is 0.28 cm in diameter and 1,250 feet long, what is the total resistance in ohms to 4 significant digits? (Be careful with units.)

   _____

7. If the power used by a component is given by the formula $P=E^2/R$, and a 100-watt lightbulb has an applied voltage of 110 volts, what is the resistance of the bulb? _____

8. If a drafter needs to set her compass to draw a circle of 4¾" diameter, what is the correct setting? _____

9. What is the mechanical advantage of a machine if the input force is 52 newtons and the output force is 239 newtons? Express your answer to the nearest tenth. _____

10. A piece of Kevlar 14" by 42" is purchased at a cost of $22.50 per square foot. What is the cost of the material? _____

11. Find the power for a ⅕ horsepower motor in watts. (1 hp = 746 W.) _____

12. What is the cutting speed in feet per minute if a ⅜" drill is turning at 1,000 rpm? _____

13. For a multi-cylinder engine, the engine displacement is calculated using the formula $0.7854 \times D^2 \times L \times$ No. of cylinders, where D is the bore diameter and L is the length of stroke. To the nearest cubic inch, what is the displacement for a 2-cylinder engine with a stroke length of 3¼" and a bore diameter of 3¼"? _____

14. If two resistors, with values of 25.4 ohms and 10.2 ohms, are connected in parallel, what is the total resistance (to the nearest tenth)? Use $R_T = \dfrac{R_1 \times R_2}{R_1 + R_2}$. _____

15. Using the same resistances as in problem 14, work the problem using the formula $\dfrac{1}{R_T} = \dfrac{1}{R_1} + \dfrac{1}{R_2}$. Compare the answers. _____

16. Using the formula from problem 6, what is the resistance in a copper wire if the diameter is 0.15 cm, the length is 1,000 m, and the resistivity of copper is 1.724×10^{-6} Ω-cm? _____

17. The coefficient of sliding friction (μ) is the ratio of the pulling force required to pull an object at a steady rate across a surface to the weight of the object. In a lab situation, a block of wood 6" long by 4" wide by 2" high was placed flat on a piece of steel. The pulling force was 8.2 N and the weight of the object was 22.8 N. Find the coefficient of sliding friction.

_____

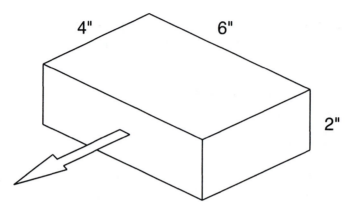

4"   6"

2"

!18. In an effort to reduce sliding friction, a suggestion was made that the object be turned on its edge rather than placed flat so that a smaller area was in contact with the surface. Based on the formula from problem 15, should that decrease the coefficient of sliding friction? Why or why not?

Geometry and Trigonometry

Geometry is used often in everyday activities and industrial technology applications, particularly in the areas of graphic communication or drafting. One important concept is the Pythagorean Theorem, which is used to find the lengths of sides of right triangles. Trigonometry is used in drafting and also in calculating the effects of forces (represented as vectors), distances, and related topics in engineering and technology applications.

Unit 41 Pythagorean Theorem

BASIC PRINCIPLES OF THE PYTHAGOREAN THEOREM

The Pythagorean Theorem is used to determine the length of a side of a *right triangle* when two of the side lengths are known. The two shorter sides of a right triangle (next to the right angle) are known as *legs*. The longest side (across from the right angle) is known as the *hypotenuse*. The terms and symbols used to refer to a right triangle are shown below.

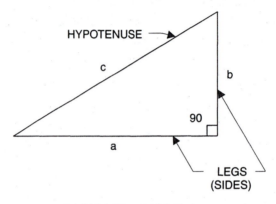

HYPOTENUSE

c

b

90

a

LEGS
(SIDES)

RIGHT TRIANGLE

The Pythagorean Theorem can be stated in words as "the sum of the squares of the legs is equal to the square of the hypotenuse." In math symbols, it is expressed as $a^2 + b^2 = c^2$

It can be written in different forms by rearranging the formula or solving it as a literal equation, as shown below.

$$c = \sqrt{a^2 + b^2} \qquad a = \sqrt{c^2 - b^2} \qquad b = \sqrt{c^2 - a^2}$$

Example: Find the length of side b in the right triangle shown.

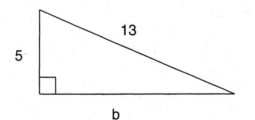

13

5

b

$$b^2 = \sqrt{c^2 - a^2}$$

$$b^2 = \sqrt{13^2 - 5^2}$$

$$b^2 = \sqrt{169 - 25} = \sqrt{144}$$

$$b = 12$$

Example: Find the length of the hypotenuse of the right triangle with sides of 5 feet and 8 feet. Round to the nearest hundredth if needed.

$$c = \sqrt{a^2 + b^2}$$

$$c = \sqrt{5^2 + 8^2}$$

$$c = \sqrt{25 + 64} = \sqrt{89}$$

$$c = 9.43'$$

Vectors

A vector is used to represent a quantity which has both a magnitude (size or value) and a direction. Some common uses of vectors are to represent a force, electrical current or voltage, velocity, and acceleration. A vector is represented graphically as a line whose length represents the magnitude and whose angle represents the direction of the quantity.

The *resultant* of vectors acting together is a single vector equivalent to the combined effect of those vectors. In this chapter, vectors which are perpendicular to each other will be studied.

Example: Find the magnitude of the resultant of vector A (35 pounds at 0°) and vector B (20 pounds at 90°).

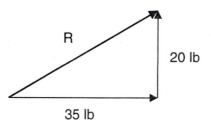

R

20 lb

35 lb

$$R = \sqrt{a^2 + b^2}$$

Use the Pythagorean Theorem, with the perpendicular vectors as the sides of the triangle.

PRACTICAL PROBLEMS

1. For each triangle listed in the chart, find the missing length of a side to the same precision as the stated values.

Right Triangles

a, b = sides   c = hypotenuse

a	b	c
1.000	1.000	
0.75	3.00	
5.0		8.6
10.0		26.0

2. If a robot gripper starts at its home position and moves 3.000" to the right and 6.000" forward, what is the straight line distance from its current position to the home position, to the nearest thousandth? Make a sketch, labeling all parts, as part of your solution. _____

3. Find the magnitude of the resultant vector (R) for vector C (25 pounds at 90°) and vector D (40 pounds at 180°). Make a sketch, labeling all parts, as part of your solution. _____

4. a. If a ramp measures 32 feet on the base and is 72" high, what is the length of the inclined surface? Make a sketch as part of your solution. Write your answer to the nearest hundredth of a foot. _____

    b. Then, without clearing the calculator, subtract the whole number part, and multiply the remaining decimal part by 12 to convert the partial foot to inches. Write the length of the inclined surface in feet and inches (to the nearest inch). _____

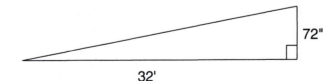

32'

5. Using the drawing, find the lengths of the inclined lines on the drawing to the nearest tenth.

x _____

y _____

z _____

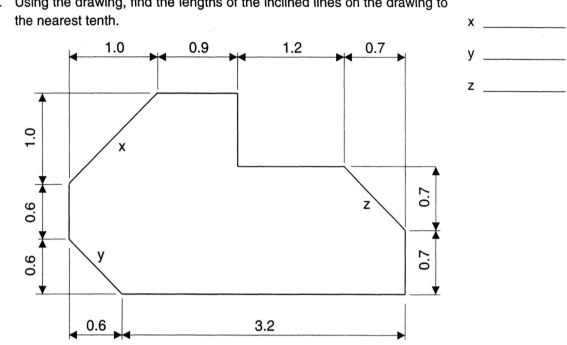

All measurements are in inches.

6. For the roof pitches listed in the table, find the length of the rafter needed to span a horizontal distance of 12'.

Roof Pitch	Horiz. run	Vert. rise	Rafter length
5/12	12'	5'	
6/12	12'	6'	
7½/12	12'	7½'	
3/12	12'	3'	

7. If a robot gripper moves 8 cm to the right and 12 cm forward from its home position, what is the straight-line distance from the home position to its current position?

_____

8. A student needs to build a ramp with a mechanical advantage of 4. If the mechanical advantage of a ramp is the same as the ratio of the length of the inclined surface to the height, what base length is needed to raise a load 4'? Round to the nearest tenth of a foot. Make a sketch as part of your solution.

_____

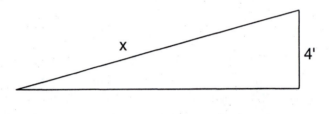

MA = 4

9. Find the magnitude of the resultant (R) of vectors A (150 N applied horizontally) and B (220 N applied vertically).

_____

10. Use the concept of similar triangles and the Pythagorean Theorem to find the length of a rafter needed to span a 16-foot horizontal run if the pitch is 5½:12.

_____

Unit 42 Trigonometric Functions

BASIC PRINCIPLES OF TRIGONOMETRIC FUNCTIONS

The word "trigonometry" is from the Greek words for "angle" and "measure." The two *legs* or sides next to the right angle are known as the *adjacent side* and the *opposite side.* The terms "adjacent" and "opposite" are used to describe a side with respect to the angle which is being considered. The longest side of a triangle is known as the *hypotenuse*, and it is always opposite the right angle. When sketching diagrams to help in solving the problems, it is helpful to use standard notation of capital letters for angles and small letters for sides of the triangle. (Angle A is typically labeled as the angle opposite side a, etc.)

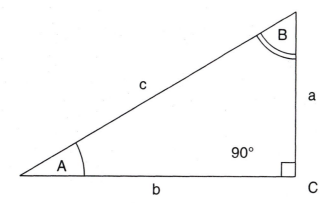

Note that in many applications, angles are identified by Greek letters such as alpha (α), beta (β), gamma (γ), or theta (θ), instead of capital letters.

Three basic trigonometric functions will be used in this unit, although six basic functions are defined. The other three are the inverses or reciprocals of the three shown below.

$$\text{sine } \theta = \sin \theta = \frac{\text{opposite side}}{\text{hypotenuse}}$$

$$\text{cosine } \theta = \cos \theta = \frac{\text{adjacent side}}{\text{hypotenuse}}$$

$$\text{tangent } \theta = \tan \theta = \frac{\text{opposite side}}{\text{adjacent side}}$$

Note that these equations can be solved for specified variables using the procedures discussed in Unit 39.

Remember that the terms *opposite* and *adjacent* are with respect to the angle being considered. The illustration below illustrates the labeling, with the angle of consideration marked θ.

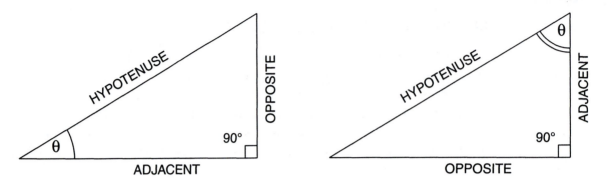

FINDING THE LENGTH OF A SIDE USING TRIG FUNCTIONS

If the length of a side and an angle of a right triangle are known, the lengths of other sides can be found using the trig functions. Setting up the equation involves identifying the sides correctly as adjacent and opposite, with respect to the known angle.

Example: Find the length of side b in the right triangle.

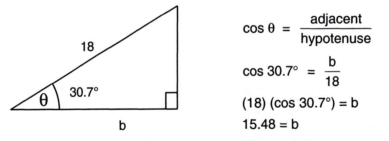

$$\cos \theta = \frac{\text{adjacent}}{\text{hypotenuse}}$$

$$\cos 30.7° = \frac{b}{18}$$

$$(18)(\cos 30.7°) = b$$

$$15.48 = b$$

FINDING AN ANGLE MEASUREMENT USING TRIG FUNCTIONS

The measurement of an angle in a right triangle can be found using trig functions. Selecting an angle in the triangle determines which side will be labeled as adjacent or opposite. When the tangent, cosine, or sine of that angle is known, the problem becomes "What angle has that cosine, sine or tangent?." The angle can be found using the calculator's inverse functions: arctangent, arccosine, or arcsin of that value. (The angles can also be identified by locating the sine, cosine or tangent value in a table and reading off the angle which matched the value. A table is provided in the Appendix of this workbook. Consult your teacher to find out whether you will use a calculator and/or trig tables.)

Example: Find the angle θ for the right triangle shown

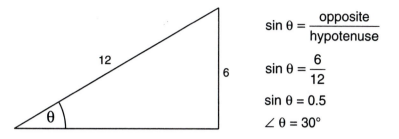

$$\sin \theta = \frac{\text{opposite}}{\text{hypotenuse}}$$

$$\sin \theta = \frac{6}{12}$$

$$\sin \theta = 0.5$$

$$\angle \theta = 30°$$

 CALCULATOR USE

Most scientific calculators are able to handle trig functions. They typically have keys such as [sin], [cos], and [tan]. On many calculators, the inverse functions [sin⁻¹], [cos⁻¹], and [tan⁻¹] are accessed by pressing a [2ndF] or [Alt] key. This key allows each key to have two different uses, with the main use printed on the key and the second or alternate use printed above or below the key.

On algebraic logic calculators, the cosine of an angle can be calculated by entering the angle value, pressing the [cos] key and then the [=] key. Sine and tangent are found using the same process. To find an angle when its cosine is known, enter the cosine value, then use the [cos⁻¹] key to display the angle.

On direct algebraic logic calculators, a cosine can be calculated by pressing the [cos] key first, entering the angle, and then pressing the [=] key. The angle whose cosine is known can be found by pressing the keys to access the [cos⁻¹] function, entering the cosine value, and pressing the [=] key. Sines and tangents are handled in the same way.

Most scientific calculators have the ability to calculate with trig functions in *degrees* or *radians*. Consult the owner's manual for your calculator to learn the specific keys or procedures to find values involving trig functions.

PRACTICAL PROBLEMS

For problems 1 and 2, identify side a, side b, and side c as adjacent, opposite, or hypotenuse, with respect to angle θ, for the right triangle shown.

1.

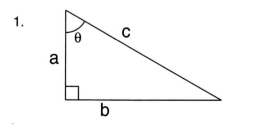

a _____

b _____

c _____

2.

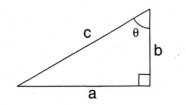

a _____

b _____

c _____

Use the illustration below for problems 3–7.

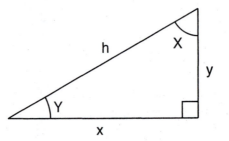

3. h = 8"   y = 6"   ∠ X =   _____

4. h = 16 cm   x = 8 cm   ∠ Y =   _____

5. y = 5 mm   x = 11 mm   ∠ X =   _____

6. y = 13"   x = 7.5"   ∠ X =   _____

7. y = 7'   h = 10'   ∠ Y =   _____

8. A roof on a house has a roof pitch of 6/12, which means that for every 12 feet of horizontal run, the roof rises 6 feet. What is the length of the rafter needed to span the run?   _____

9. A ramp is to be built with a maximum angle of 12 degrees from horizontal. If the height of the ramp is to be 4 feet, what is the minimum length for the base?   _____

10. A gable roof on a warehouse was built at a 3/12 pitch. If the building is 80 feet wide, what is the height of the roof at the center of the building? Make a sketch as part of your solution.   _____

11. A cover must be designed to fit on the back of an antenna base to cover the installation screws. Using the dimensions in the illustration, find the length (x) of the cover. Make a sketch of the triangle represented in the problem and label the parts as a part of your solution.

_____

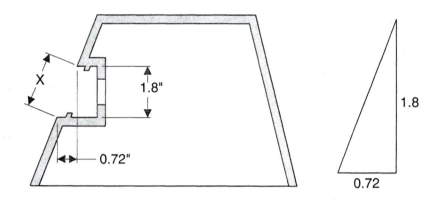

12. The 3/4/5 triangle and the 5/12/13 triangle are often called Pythagorean Triples because they fit the Pythagorean theorem with whole number values for all three sides. For each triangle, find the angle measurements for the angles marked A, B, C, and D.

A _____

B _____

C _____

D _____

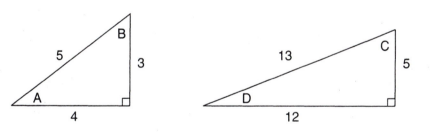

EXAMPLES OF PYTHAGOREAN TRIPLES

13. A pipe fitter must bend a pipe to fit between the two pipes as shown in the diagram. Find angle θ and length x.

θ _____

x _____

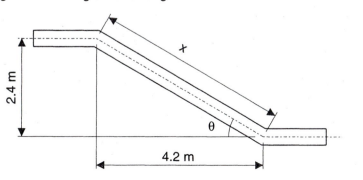

!14. Use a calculator (or convert radians to degrees and use a trig table) to find the following values, given the angles stated in radians:

    a. $\cos 2\pi$ _____

    b. $\cos \pi/2$ _____

    c. $\tan \pi/6$ _____

    d. $\sin 3\pi/4$ _____

Appendix

RESISTOR COLOR CODE

Color Code	1st band 1st Digit	2nd band 2nd Digit	3rd band Power of 10	4th band Tolerance
0 Black	0	0	0	
1 Brown	1	1	1	
2 Red	2	2	2	
3 Orange	3	3	3	
4 Yellow	4	4	4	
5 Green	5	5	5	
6 Blue	6	6	6	
7 Violet	7	7	7	
8 Gray	8	8	8	
9 White	9	9	9	
Gold				5%
Silver				10%
No Color				20%

U.S. CUSTOMARY SYSTEM MEASUREMENT EQUIVALENTS

Length

1 foot (ft or ') = 12 inches (in or ")
1 yard (yd) = 3 feet (ft or ')
1 mile (mi) = 1,760 yards (yd)
1 mile (mi) = 5,280 feet (ft or ')

Area

1 square yard (sq yd or yd^2) = 9 square feet (sq ft or ft^2)
1 square foot (sq ft or ft^2) = 144 square inches (sq in or in^2)

Volume for Solids

1 cubic yard (cu yd or yd^3) = 27 cubic feet (cu ft or ft^3)
1 cubic foot (cu ft or ft^3) = 1,728 cubic inches (cu in or in^3)

Volume for Fluids

1 gallon (gal) = 4 quarts (qt)
1 gallon (gal) = 0.133681 cubic foot (cu ft)
1 gallon (gal) = 231 cubic inches (cu in)

Other Units Used in the U.S. Customary System

Acceleration: feet per second squared (ft/s^2)
Energy: foot-pound (ft-lb)
Force: pound (lb) or ounce (oz)
Mass: ounce mass (oz mass) or pound mass (lb mass)
Moment of a force: pound-foot (lb-ft) or pound-inch (lb-in)
Power: foot-pound per second (ft-lb/s) or horsepower (hp)
Pressure: pounds per square foot (lb/ft^2) or pounds per square inch (lb/in^2)
Velocity: feet per second (ft/s) or miles per hour (mph or mi/h)
Work: foot-pound (ft-lb)

SI (METRIC) SYSTEM MEASUREMENT EQUIVALENTS

Length

1 centimeter (cm) = 10 millimeters (mm)
1 meter (m) = 100 centimeters (cm) = 1,000 millimeters (mm)
1 kilometer (km) = 1,000 meters (m)

Area

1 square centimeter (cm^2) = 100 square millimeters (mm^2)
1 square meter (m^2) = 10,000 square centimeters (cm^2) = 10^4 cm^2
1 square meter (m^2) = 1,000,000 square millimeters (mm^2) = 10^6 mm^2

Volume for Solids

1 cubic meter (m^3) = 10^9 cubic millimeters (mm^3)
1 cubic meter (m^3) = 10^6 cubic centimeters (cm^3)

Volume for Fluids

1 liter (L) = 1,000 milliliters (mL)
1 liter (L) = 1,000 cubic centimeters (cm^3)
1 milliliter (mL) = 1 cubic centimeter (cm^3 or cc)

Mass

1 gram (g) = 1,000 milligrams (mg)
1 kilogram (kg) = 1,000 grams (g)

Other Units Used in the SI System

Acceleration: meter per second squared (m/s^2)
Energy: joule (J) = 1 N-m
Force: newton (N) = 1 kg-m/s^2
Moment of a force: newton-meter (N-m)
Power: Watt (W) = 1 J/s
Pressure: Pascal (Pa) = 1 N/m^2
Velocity: meter per second (m/s)
Work: joule (J) = 1 N-m

U.S. - METRIC EQUIVALENTS

Length

1 inch (in) = 2.54 centimeters (cm) = 25.4 millimeters (mm)
1 foot (ft) = 0.3048 meter (m)
1 yard (yd) = 0.9144 meter (m)
1 mile (mi) = 1.609 kilometers (km)

Area

1 square inch (in^2) = 6.4516 square centimeters (cm^2)
1 square foot (ft^2) = 0.092903 square meter (m^2)
1 square yard (yd^2) = 0.836127 square meter (m^2)

Volume for Solids

1 cubic inch (in^3) = 16.387064 cubic centimeters (cm^3)
1 cubic foot (ft^3) = 0.028317 cubic meter (m^3)
1 cubic yard (yd^3) = 0.764555 cubic meter (m^3)

Volume for Fluids

1 gallon (gal) = 3.785411 liters (L)
1 quart (qt) = 0.946353 liter (L)
1 ounce (oz) = 29.573530 cubic centimeters (cm^3 or cc)

Mass

1 pound (lb) = 0.453592 kilogram (kg)
1 pound (lb) = 453.59237 grams (g)
1 ounce (oz) = 28.349523 grams (g)

Other Units

Acceleration: 1 ft/s^2 = 0.3048 m/s^2
Angle measure: 180° = π radians
Energy: 1 ft-lb = 1.356 J
Force: 1 lb = 4.448 N
Moment of a force: 1 lb-ft = 1.356 N-m
Power: 1 ft-lb/s = 1.356 W or 1 hp = 745.7 W
Pressure: 1 lb/ft^2 = 47.88 Pa or 1 lb/in^2 (psi) = 6.895 kPa
Velocity: 1 ft/s = 0.3048 m/s   or 1 mi/h (mph) = 1.609 km/h
Work: 1 ft-lb = 1.356 J

FORMULAS

Electricity

Resistances in series: $R_T = R_1 + R_2 + \ldots + R_n$

Resistances in parallel: $1/R_T = 1/R_1 + 1/R_2 + \ldots + 1/R_n$

$E = I\,R$

$I = E/R$ where E = voltage in volts (v)

$R = E/I$ R = resistance in ohms (Ω)

$P = I\,E$ I = amperage in amperes (A)

$P = I^2\,R$ P = power in watts (W)

Temperature

$^\circ C = \frac{5}{9}\,(^\circ F - 32)$

$^\circ F = \frac{9}{5}\,^\circ C + 32$

Area, Volume, and Circumference

$A = \pi r^2$ where A = area, r = radius of a circle

$A = (L)(W)$ where A = area of a rectangle, L = length, and W = width

$A = (1/2)bh$ where A = area of a triangle, b = base, and h = height

$V = \pi r^2\,h$ where V = volume of a cylinder, r = radius, and h = height

$V = (L)(W)(H)$ where V = volume of a rectangular solid, L = length, W = width, H = height

$C = 2\pi r$ or πd where C = circumference of a circle, r = radius, and d = diameter

Other Formulas Used in Industrial Technology

$\text{rpm} = \dfrac{CS' \times (12''/1')}{d''\,(\pi)}$ where CS is cutting speed in ft/min and d is diameter

$D = M \div V$ where D = density, M = mass, and V = volume

$P = F \div A$ where P = pressure, F = force, and A = area

$k = F \div \Delta L$ where k = spring constant, F = force and ΔL = change in length

$I = P\,r\,t$ where I = interest, P = principal, r = annual interest rate, and t = time

$ma = F_0 \div F_i$ where ma = mechanical advantage, F_0 = output force, and F_i = input force

$m = \text{rise/run}$ where m = slope, rise = vertical change, run = horizontal change

$\text{Bd ft} = \dfrac{(\text{nom. size})(L'')}{144}$ where Bd ft = board feet, L = length (")

$\text{Bd ft} = \dfrac{(\text{nom. size})(L')}{12}$ where Bd ft = board feet, L = length (')

Other Formulas Used in Industrial Technology (Cont'd)

$\%e = \Delta L \div L_o$ where %e = % elongation, ΔL = change in length, and L_o = original length

$KE = (1/2)mv^2$ where KE = kinetic energy, m = mass, and v = velocity

$W = F\,D$ where W = work, F = force, D = distance

$v = \Delta d/t$ where v = velocity, Δd = change in distance, and t = time

$a = \Delta v/t$ where a = acceleration, Δv = change in velocity, and t = time

$a^2 + b^2 = c^2$ where a, b = sides of right triangle, and c = hypotenuse (Pythagorean Theorem)

Trigonometry Formulas

sine of an angle (sin) = $\dfrac{\text{opposite side}}{\text{hypotenuse}}$

cosine of an angle (cos) = $\dfrac{\text{adjacent side}}{\text{hypotenuse}}$

tangent of an angle (tan) = $\dfrac{\text{opposite side}}{\text{adjacent side}}$

TRIGONOMETRIC FUNCTIONS

Angle	Sine	Cosine	Tangent	Cotangent	Secant	Cosecant
1	0.0175	0.9998	0.0175	57.2900	1.0002	57.2987
2	0.0349	0.9994	0.0349	28.6363	1.0006	28.6537
3	0.0523	0.9986	0.0524	19.0811	1.0014	19.1073
4	0.0698	0.9976	0.0699	14.3007	1.0024	14.3356
5	0.0872	0.9962	0.0875	11.4301	1.0038	11.4737
5 5/8	0.0980	0.9952	0.0985	10.1532	1.0048	10.2023
6	0.1045	0.9945	0.1051	9.5144	1.0055	9.5668
7	0.1219	0.9925	0.1228	8.1443	1.0075	8.2055
8	0.1392	0.9903	0.1405	7.1154	1.0098	7.1853
9	0.1564	0.9877	0.1584	6.3138	1.0125	6.3925
10	0.1736	0.9848	0.1763	5.6713	1.0154	5.7588
11	0.1908	0.9816	0.1944	5.1446	1.0187	5.2408
11 1/4	0.1951	0.9808	0.1989	5.0273	1.0196	5.1258
12	0.2079	0.9781	0.2126	4.7046	1.0223	4.8097
13	0.2250	0.9744	0.2309	4.3315	1.0263	4.4454
14	0.2419	0.9703	0.2493	4.0108	1.0306	4.1336
15	0.2588	0.9659	0.2679	3.7321	1,0353	3.8637
16	0.2756	0.9613	0.2867	3.4874	1.0403	3.6280
17	0.2924	0.9563	0.3057	3.2709	1.0457	3.4203
18	0.3090	0.9511	0.3249	3.0777	1.0515	3.2361
19	0.3256	0.9455	0.3443	2.9042	1.0576	3.0716
20	0.3420	0.9397	0.3640	2.7475	1.0642	2.9238
21	0.3584	0.9336	0.3839	2.6051	1.0711	2.7904
22	0.3746	0.9272	0.4040	2.4751	1.0785	2.6695
22 1/2	0.3827	0.9239	0.4142	2.4142	1.0824	2.6131
23	0.3907	0.9205	0.4245	2.3559	1.0864	2.5593
24	0.4067	0.9135	0.4452	2.2460	1.0946	2.4586
25	0.4226	0.9063	0.4663	2.1445	1.1034	2.3662
26	0.4384	0.8988	0.4877	2.0503	1.1126	2.2812
27	0.4540	0.8910	0.5095	1.9626	1.1223	2.2027
28	0.4695	0.8829	0.5317	1.8807	1.1326	2.1301
29	0.4848	0.8746	0.5543	1.8040	1.1434	2.0627
30	0.5000	0.8660	0.5774	1.7321	1.1547	2.0000
31	0.5150	0.8572	0.6009	1.6643	1.1666	1.9416
32	0.5299	0.8480	0.6249	1.6003	1.1792	1.8871
33	0.5446	0.8737	0.6494	1.5399	1.1924	1.8361

Angle	Sine	Cosine	Tangent	Cotangent	Secant	Cosecant
34	0.5592	0.8290	0.6745	1.4826	1.2062	1.7883
35	0.5736	0.8192	0.7002	1.4281	1.2208	1.7434
36	0.5878	0.8090	0.7265	1.3764	1.2361	1.7013
37	0.6018	0.7986	0.7536	1.3270	1.2521	1.6616
38	0.6157	0.7880	0.7813	1.2799	1.2690	1.6243
39	0.6293	0.7771	0.9098	1.2349	2.2868	1.5890
40	0.6428	0.7660	0.8391	1.1918	1.3054	1.5557
41	0.6561	0.7547	0.8693	1.1504	1.3250	1.5243
42	0.6691	0.7431	0.9004	1.1106	1.3456	1.4945
43	0.6820	0.7314	0.9325	1.0724	1.3673	1.4663
44	0.6947	0.7193	0.9657	1.0355	1.3902	1.4396
45	0.7071	0.7071	1.0000	1.0000	1.4142	1.4142
46	0.7193	0.6947	1.0355	0.9657	1.4396	1.3902
47	0.7314	0.6820	1.0724	0.9325	1.4663	1.3673
48	0.7431	0.6691	1.1106	0.9004	1.4945	1.3456
49	0.7547	0.6561	1.1504	0.8693	1.5243	1.3250
50	0.7660	0.6428	1.1918	0.8391	1.5557	1.3054
51	0.7771	0.6293	1.2349	0.9098	1.5890	1,2868
52	0.7880	0.6157	1.2799	0.7813	1.6243	1.2690
53	0.7986	0.6018	1.3270	0.7536	1.6616	1.2521
54	0.8090	0.5878	1.3764	0.7265	1.7013	1.2361
55	0.8192	0.5736	1,4281	0.7002	1.7434	1.2208
56	0.8290	0.5592	1.4826	0.6745	1.7883	1.2062
57	0.8387	0.5446	1.5399	0.6494	1.8361	1.1924
58	0.8480	0.5299	1.6003	0.6249	1.8871	1.1792
59	0.8572	0.5150	1.6643	0.6009	1.9416	1.1666
60	0.8660	0.5000	1.7321	0.5774	2.0000	1.1547
61	0.8746	0.4848	1.8040	0.5543	2.0627	1.1434
62	0.8829	0.4695	1.8807	0.5317	2.1301	1.1326
63	0.8910	0.4540	1.9626	0.5095	2.2027	1.1223
64	0.8988	0.4384	2.0503	0.4877	2.2812	1.1126
65	0.9063	0.4226	2.1445	0.4663	2.3662	1.1034
66	0.9135	0.4067	2.2460	0.4452	2.4586	1.0946
67	0.9205	0.3907	2.3559	0.4245	2.5593	1.0864
68	0.9272	0.3746	2.4751	0.4040	2.6695	1.0785
69	0.9336	0.3584	1.6051	0.3839	2.7904	1.0711
70	0.9397	0.3420	2.7475	0.3640	2.9238	1.0642
71	0.9455	0.3256	2.9042	0.3443	3.1916	1.0576

Angle	Sine	Cosine	Tangent	Cotangent	Secant	Cosecant
72	0.9511	0.3090	3.0777	0.3249	3.2361	1.0515
73	0.9563	0.2924	3.2709	0.3057	3.4203	1.0457
74	0.9613	0.2756	3.4874	0.2867	3.6280	1.0403
75	0.9659	0.2588	3.7321	0.2679	3.8637	1.0353
76	0.9703	0.2419	4.0108	0.2493	4.1336	1.0306
77	0.9744	0.2250	4.3315	0.2309	4.4454	1.0263
78	0.9781	0.2079	4.7046	0.2126	4.8097	1.0223
79	0.9816	0.1908	5.1446	0.1944	5.2408	1.0187
80	0.9848	0.1736	5.6713	0.1763	5.7588	1.0154
81	0.9877	0.1564	6.3138	0.1584	6.3925	1.0125
82	0.9903	0.1392	7.1154	0.1405	7.1853	1.0098
83	0.9925	0.1219	8.1443	0.1228	8.2055	1.0075
84	0.9945	0.1045	9.5144	0.1051	9.5668	1.0055
85	0.9962	0.0872	11.4301	0.0875	11.4737	1.0038
86	0.9976	0.0698	14.3007	0.0699	14.3356	1.0024
87	0.9986	0.0523	19.0811	0.0524	19.1073	1.0014
88	0.9994	0.0349	28.6363	0.0349	28.6537	1.0006
89	0.9998	0.0175	57.2900	0.0175	57.2987	1.0002

Glossary

Acceleration - For an object in motion, the change in velocity divided by the change in time.

Alternating current - An electric current that changes direction at regular intervals.

Ampere - The unit of electrical current, a rate of flow equal to 1 coulomb per second.

Architect's scale - A rule, typically with ten different scales divided into feet and inches, and one with $\frac{1}{16}$" increments.

Area - A measure of the size of a two-dimensional surface or plane.

Board foot - A standard measure of lumber based on the nominal size and equal to 144 in^3.

British thermal unit (Btu) - The amount of heat needed to raise the temperature of one pound of water by 1°F.

Bushing - A cylindrical metal sleeve used in machines to reduce friction or decrease an inside diameter.

Cartesian coordinate system - A grid system with perpendicular axes in which each location is named by an ordered pair (x,y).

Circuit - A path through which electrical current can flow.

Circuit breaker - A device that automatically opens a circuit when the current exceeds a preset value or when the voltage is excessive; it does the same job as a fuse, but can be reset.

Civil engineer's scale - A rule having several scales, divided so that the number of increments per inch is a multiple of ten.

Coordinate - An x or y value in an ordered pair.

Coulomb - The metric unit of electrical charge, it is equal to the combined charge of 6.24×10^{18} electrons.

Cross-sectional area - An area measurement for a surface obtained by cutting an object with a cutting plane, often perpendicular to the length on an object such as wire.

Current - The flow of electrons through a circuit.

Density - The mass of a quantity of material divided by its volume.

Direct current - Electrical current which flows continuously in the same direction.

Discount - The amount of reduction in the price of an item, normally stated as a percentage of the list price.

Efficiency - The useful output of a device or system divided by the input.

Electricity - In practical applications, it is the use of electrical current to provide power for such devices as lights and motors.

Electrical resistance - A measure of the opposition to electrical current flow exhibited by a component or circuit, it is measured in ohms.

Energy - The capacity for doing work.

Fossil fuel - Fuel such as coal or types of petroleum which are normally burned to provide heat energy.

Gross income - Income before taxes, retirement, insurance, or other deductions are subtracted .

Horsepower - A unit of power, it is equal to 746 watts.

Joule - A metric unit of work or energy, it is the work done by a force of one newton moving an object one meter.

Kilowatt-hour (kWh) - The typical unit of energy by which electrical energy is sold to consumers, it is equal to a power of 1,000 watts used over a period of 1 hour.

Kinetic energy - The energy in a moving object due to its motion, it is defined as one-half of the object's mass multiplied by the square of its velocity.

Limits - The maximum and minimum values for a dimension on a part.

Micrometer - A device used to measure distances, it is available for both inch and metric measurements.

Mass - Often described as the amount of material in an object, it is a measure of the object's resistance to acceleration.

Net income - Income after taxes, retirement, insurance and other deductions are subtracted.

Newton (N) - The metric unit of force, it is the force required to accelerate one kilogram at one meter per second for each second.

Nominal dimension - The designated size of an object, which may be different from the actual size (for example, 2 × 4 lumber) .

Ohm - A measure of resistance to electrical current through a component or circuit.

Ohm's Law - A statement of the relationship between voltage, amperage and resistance. It can be stated as E/R = I, E/I = R, or I × R = E where E is voltage, R is resistance, and I is current.

Origin - A reference point on the Cartesian coordinate system, having the ordered pair (0,0).

Parallel circuit - A circuit with more than one path for current flow.

Perimeter - The distance around the edge of a surface.

Potential energy - The energy of an object due to its position, it is energy that can be converted to another form of energy, such as kinetic energy.

Power - The rate of doing work or expending energy, often measured in watts.

Pressure - The force applied to a surface divided by the area of the surface over which the force acts.

Principal - The amount of money invested or borrowed.

Pythagorean Theorem - A theorem in geometry which states that the square of the hypotenuse in a right triangle is equal to the sum of the squares of the other two sides.

RPM (revolutions per minute) - A measure of rotary motion, the number of complete (360°) rotations of an object around an axis in one minute.

R-value - A measure of the resistance to heat transfer through a material.

Resistor - A component used to increase the electrical resistance in a circuit.

Sale price - The price of an item after any discounts are subtracted.

Scale - (1) An instrument used to measure; (2) a proportion between the sizes of objects, such as a drawing and the actual object.

Scientific notation - A mathematical representation of a value in which a value between 1 and 10 is multiplied by the appropriate power of 10.

Series circuit - A circuit which has one path for current flow through components.

Solder - (1) a joining process in which a filler material is melted and flows to fill a joint; (2) the filler material commonly used in such a process.

Speed - The change in distance divided by the change in time.

Stress - An internal reaction in a material to an applied force; measured in the same units as pressure.

Tolerance - The total amount of variation allowed in the size or dimension of a part, it is the difference between the maximum and minimum dimensions or limits.

Velocity - For an object in motion, the change in distance divided by the change in time, and including the direction of motion.

Vernier - A modification of a standard measuring device such as a micrometer which allows measurement to be made to an additional decimal place or degree of accuracy.

Volt - Unit of electrical force or potential difference.

Volume - A measure of the size of a three-dimensional space or object.

Watt - Metric unit of power, it is a rate of doing work of one joule per second.

Weight - The mass of an object multiplied by the effect of the gravitational force.

Work - The product of the distance an object is moved times the force acting in the direction that the object moves.

Answers to Odd-Numbered Problems

Unit 1: Addition of Whole Numbers

1. 82
3. 772
5. 3,964
7. 22,121
9. 4,078

11. 18
13. 173 lb
15. 155 parts
17. a. $1007
    b. yes

19. 125,920 gallons
21. 582 gallons
23. 83'
25. $562

Unit 2: Subtraction of Whole Numbers

1. 314
3. 633
5. 110,946
7. 13,755
9. 8,642

11. 38
13. 630 miles
15. 23 minutes
17. 913° F

19. a. 35 (3½")
    b. 18 (5¼")
21. $198
23. 191 boxes
25. $286 profit

Unit 3: Multiplication of Whole Numbers

1. 444
3. 44,394
5. 667,116
7. 4,049
9. 8,384,234
11. 164,250 rev

13. 2,200 turns
15. $24,000
17. 120 volts
19. 70,875 feet
21. $432

23. 180 ft-lb
25. 103,680
27. 48,400
29. $1,428
31. a. 3,062 kWh
    b. $287.70

Unit 4: Division of Whole Numbers

1. 308
3. 2,712
5. 30
7. $2 \cdot 2 \cdot 2 \cdot 2 \cdot 2 \cdot 2$
9. 87

11. 14 crates
13. 85 gross
15. 14 days
17. a. 3" horiz.
    b. 2" vert.

19. a. 174 @ $5
    b. 58 @ $15
21. 54 bicycles
23. 97 N/m^2
25. 48 minutes

Unit 5: Review and Combined Operations on Whole Numbers

1. 898,656
3. 18
5. 70
7. $5 \cdot 5 \cdot 7$
9. 92,274
11. $516
13. a. 171 L
   b. 79 L

15. a. 363 bundles
   b. 121 squares
17. a. $832 tuition for 16 hours
   b. $1,102 total
   c. $208 for 4 hr. tuition
19. a. no
   b. 2 more 12' pieces

21. $198
23. 45 cartons
25. 144
27. a. 3" horiz
   b. 2" vert
29. $3540

Section 3: Common Fractions

1. $\frac{2}{3} = \frac{4}{6} = \frac{12}{18} = \frac{14}{21}$
3. $\frac{3}{5} = \frac{6}{10} = \frac{15}{16} = \frac{24}{40}$
5. $\frac{5}{16}$
7. $\frac{3}{4}$
9. $\frac{3}{1}$; $\frac{12}{1}$; $\frac{8}{1}$; $\frac{15}{1}$

11. $\frac{2}{16}$; $\frac{8}{16}$
13. $\frac{10}{15}$; $\frac{12}{15}$
15. $\frac{11}{64}$; $\frac{26}{64}$
17. $6\frac{7}{8}$

19. $15\frac{3}{16}$
21. $\frac{37}{5}$
23. $\frac{171}{8}$
25. $\frac{175}{8}$

Unit 6: Addition of Common Fractions

1. $\frac{1}{3}$
3. $\frac{27}{64}$
5. $\frac{251}{150}$
7. $31\frac{5}{8}$
9. $13\frac{19}{64}$
11. $1\frac{3}{8}$"

13. $6\frac{1}{2}$"
15. a. $5\frac{1}{2}$ hours
   b. under
   c. $\frac{3}{4}$ hour
17. $9\frac{1}{2}$"

19. $8\frac{1}{2}$" $\times$ $7\frac{7}{8}$"
21. $2\frac{13}{120}$
23. 538¢; $5.38
25. $1\frac{23}{32}$"

Unit 7: Subtraction of Common Fractions

1. $\frac{1}{4}$
3. $\frac{15}{64}$
5. $\frac{373}{800}$
7. $16\frac{1}{8}$
9. $5\frac{9}{16}$
11. a. $1\frac{3}{4}$
   b. $2\frac{1}{4}$

13. $\frac{5}{8}$"
15. a. max $6\frac{1}{4}$"
      min $6\frac{1}{8}$"
   b. tol $\frac{1}{8}$"
17. $1\frac{13}{16}$"
19. $\frac{3}{16}$"
21. $6\frac{5}{8}$"

23. $33\frac{1}{4}$"
25. A = $\frac{11}{16}$"
   B = $\frac{1}{4}$"
   C = $\frac{1}{2}$"
   D = $1\frac{1}{2}$"

Unit 8: Multiplication of Common Fractions

1. $3/16$
3. $9/64$
5. $9\frac{5}{8}$
7. $25/3 = 8\frac{1}{3}$
9. $25/64$

11. a. $4(5\frac{1}{2})$ ohms
     b. 22 ohms
13. $30
15. $44\frac{7}{8}$"
17. $25/32$"
19. $146\frac{1}{4}$ lb

21. a. $3/8$ lb/part
     b. 240 lb
23. $46\frac{3}{4}$"
25. $15/16$ ft-lb
27. L=$2\frac{13}{16}$"
     W=$1\frac{3}{4}$'
     H=$1\frac{1}{8}$'

Unit 9: Division of Common Fractions

1. $5/6$
3. $4\frac{15}{16}$
5. $1\frac{3}{11}$
7. $9\frac{1}{8}$

9. $347/448$
11. $18\frac{10}{19}$
13. 12

15. x = $1\frac{1}{2}$" horiz.
     y = $3/4$" vert.
17. 4 A
19. $\frac{1}{2}$"

Unit 10: Review and Combined Operations on Common Fractions

1. $1\frac{9}{16}$
3. $1\frac{9}{32}$
5. $5/8 = 10/16 = 20/32 = 35/56$
7. $1\frac{43}{192}$
9. $1\frac{1}{8}$
11. A = $15/16$
     B = $1\frac{11}{16}$
     C = $2\frac{1}{4}$

13. A = $\frac{1}{4}$
     B = $5/8$
     C = $1\frac{1}{4}$
     D = $1\frac{7}{8}$
     A + C = $1\frac{1}{2}$
     D − B = $1\frac{1}{4}$
     D − (A + B) = 1
15. 45
17. $3/16$ inch
19. 129 hours

21. $3/32$"
23. $7/32$ inch
25. A = $5/8$"
     B = $4\frac{7}{16}$"
     C = $1\frac{1}{16}$"
27. 22 pieces
29. A = $7/8$"
     B = $1\frac{3}{16}$"
     C = $9/16$"
     D = $5/8$"

Unit 11: Significant Digits, Rounding, and Scientific Notation

1. 7.4
3. 21.4
5. 548.313
7. 27.38
9. 12.49

11. 32.15 ± 0.01
     32.100 ± 0.005
     25.325 ± 0.005
     27.30 ± 0.01
13. 50.1

15. 1.4
17. 38,500
19. .005625
21. 3.6×10^6 J
23. .049 in^2

Unit 12: Addition of Decimal Fractions

1. 8.7
3. 1387.11
5. 93.589
7. 1.3255
9. 2,122.755

11. 0.013"
13. $29.34
15. $12.25
17. $ 0.54

19. 4.086 A
21. 4.545 A
23. 6.150 x 10^9 W or
     6,150,000,000 W

Unit 13: Subtraction of Decimal Fractions

1. 5.1
3. 106.625
5. .030482
7. 0.7495
9. 16.325
11. 0.085"
13. a. $196.80 cost
     b. $68.20 profit

15. A=0.4
     B=3.40
     C=1.475
     D=.375
17. 0.035"
19. 10.27" min and 10.33" max
     2.371" min and 2.379" max
     1.87" min and 1.89" max
     4.621" min and 4.629" max

0.44" min and 0.46" max
6.471" min and 6.429" max
2.94" min and 2.96" max
21. A = 0.708
     B = 0.325
     C = 0.488
     D = 1.508
23. 0.7875"

Unit 14: Multiplication of Decimal Fractions

1. 6.3
3. 0.451665
5. 8.125
7. .047912
9. 0.36
11. 22.2 v
13. 3,750,000,000 W =
     3.75 × 10^9 W

15. a. total earnings $534.60
     b. income tax $64.15
     c. SS $44.64
     d. net pay $425.81
17. a. $296.00
     b. $10.73
     c. $343.36

19. $0.75
21. a. $383.40
     b. $69.01
23. a. 300 lb
     b. 9,000,000 = 9 × 10^6 lb
25. 12; 18

Unit 15: Division of Decimal Fractions

1. 21.95
3. 14.74
5. 1.832 x 10^2 or 183.2
7. 95.3
9. 62.5

11. 29.2 mi/gal
13. 10.5 A
15. 0.525"
17. 25¢

19. 0.035
21. 2.46 s
23. 11.4 lb/in
25. a. 30 pieces
     b. 0.6" left

Unit 16: Decimal and Common Fraction Equivalents

1. 0.625
3. 2.171875
5. 2/5
7. 2 1/40
9. a. 5/16
    b. 4/16 = 0.25, and
        6/16 = 0.375
    c. 0.25 < 0.333 < 0.375
11. 7/8

13. 3/8" = 0.375"
15. 10.55" = 10 9/16"
    10.63" = 10 5/8"
    27.00" = 27"
    15.70" = 15 11/16"
    10.90" = 10 7/8"
    37.50" = 37 1/2"
17. 0.880"
19. 1.631"

21. 5/16 = 0.3125
23. 2/3 = 0.667
    3/4 = 0.75
    5/6 = 0.833
    3/8 = 0.375
    5/8 = 0.625
    7/8 = 0.875
    7/16 = 0.4375

Unit 17: Review and Combined Operations on Decimal Fractions

1. 29.823
3. 15.4275
5. .052854
7. 3.6395×10^3
9. 10,474.31
11. 0.013"
13. a. min. clear. = 0.029
    b. max. clear. = 0.035

15. 0.125
17. I = 0.624 A
    P = 4.867 W
19. $149.26
21. Taxes: $49.82
    Retirement $24.91
    Net: $ 236.67

23. $84.70
25. $19,662.50
27. a. reduction 0.012"
    b. offset 0.006"
29. $556.05

Unit 18: Ratios

1. 6/1
3. 1.8/1 or 18/10
5. 29/150
7. 1/4.3
9. yes
11. a. 16/1
    b. 16/17
13. 3/1

15. a. 15/11
    b. 9/1
    c. 26/7
17. a. 3/10
    b. 7/10
    c. 3/7
19. a. 3/50
    b. 3/47

21. 8/1
23. 11/2
25. 5/12
27. a. 1/2
    b. 1/2
    c. equal
29. no

Unit 19: Proportion

1. x = 16
3. x = 5.4
5. x = 2
7. x = 35
9. 16,510
11. 80
13. a. 3.2 lb tin
    b. 35.2 lb bronze

15. chem: 7 oz
    water: 14 oz
17. 0.71
19. A = 1/2
    B = 4/7
    C = 2/5
    D = 1
21. 0.0203 ohms

23. a. ¼" : 1' :: x : 82'
    b. 20.5" long
    c. 11.75" wide
25. 7/12
27. a. 37.5 psi
    b. 25 psi
29. 540 s

Unit 20: Combined Problems in Ratio and Proportion

1. 5/16
3. 7:450
5. yes
7. 504
9. 24
11. 2.8

13. 0.3 L conc.
    21.7 L water
15. a. 29.25 lb copper
        15.75 lb zinc
    b. 780 lb copper
        420 lb zinc

17. a. 1/16
    b. 1/17
    c. 10 oz chemical
        160 oz water
19. a. 5/3
    b. 1,440 rpm
21. 8/1

Unit 21: Percent and Percentages

1. 88%
3. 0.74
5. 87.5%
7. 2⅕
9. ⅝
11. 18%
13. 3%
15. 62.5%

17. 8.8%
19. 59.375%
21. 82.6%
23. 92%
25. First: $6400
    Second: $10,240
27. $43.82

29. a. 30%
    b. 99 lb copper
        27 lb nickel
        54 lb zinc
31. $ 36.94
33. 0.007
    0.0015
35. 13.7%

Unit 22: Simple Interest

1. $432; $2,232
3. $153,000;
    $ 238,000
5. 0.000411
    $ 40.07
7. 0.0075
    $67.50
9. $3,323.20

11. $13,600
13. a. 0.00833
     b. $129.95
15 $300.00
17. On 6/1: 2 months
        $324.00 int
        $41,824.00 new
        balance

On 8/1: 2 months
    $752.83 int
    $74,576.83 new
    balance
On 9/1: 1 month
    $671.19 int
    $75,248.02 payoff

Unit 23:Discount

1. $330.00
3. $79.00
5. $20,074.05
7. $92.50
    $1,757.50
    $123.02
    $1,634.48

9. $126.60
    $928.40
    $37.14
    $891.26
    $17.83
    $873.43
11. 225.29

13. a. $391.30
     b. 45.5%
15. $80
     $920
     $55.20
     $864.80
     $25.94
     $838.86
     answers should match

Unit 24: Averages and Estimates

1. 10×100=1,000 est.
    1,164 actual
    164 difference
    (estimate may vary)
3. 2,000×40=80,000 est
    83,250 actual
    3,250 difference
    (estimate may vary)

5. 3,000×2,000=6,000,000
    6,080,000 actual
    80,000 diff.
    (estimate may vary)
7. 16,663
9. 41.19
11. 30.8

13. a. 2×10 = 20 hrs. est
     b. 20 hr. × 20 $/hr
        = $400 est.
     $378.00 actual
15. a. $3,690 allowance
     b. $3,835 cost
     c. 4% difference
17. 5,600 hours
19. 7.9%

Unit 25: Review of Problems Involving Percents, Averages, and Estimates

1. 31.25%
3. 1¾
5. 260
7. $264 int
   $2,464 payoff

9. $76.46
11. 3%
13. $1,872
15. 9.25%
17. 55%

19. $918.54
21. a. 119.6 volts
    b. 128 and 112 are
       out of spec

Unit 26: Exponents and Order of Operations with Exponents

1. 125
3. 18
5. 1,849
7. 113
9. 0.000001724 ohm-cm

11. 66.15 W
13. a. 0.0082 in^2
    b. 0.0327 in^2
    c. 4:1
    d. 4

15. 2,694 W
17. 117
19. 107

Unit 27: Roots

1. 8
3. 16.5
5. 9

7. 10
9. 4.8
11. 10.41

13. 7.1
15. 249.7 v.
17. 12.5'

Unit 28: Review and Combined Operations on Exponents and Roots

1. 5,329
3. 23
5. 13

7. 5
9. 13
11. 79

13. 13.849"
15. 1,625 kg-m^2/s^2
17. 184.632 m^3

Unit 29: Length and Angle Measurement

1. 250 cm
3. 425 mm
5. 4.19 radians
7. 225°
9. 43.4 km
11. 8'

13. a. 741 mi/hr
    b. 662.6 m/s
15. 0.4
    12.7; 0.8
17. a. ¼" : 1'
    b. 21.5" long
    c. 10.5" wide

19. 63"
21. A = 0.7
    B = 1.3
    C = 1.8
    D = 2.6

Unit 30: Area and Pressure Measurement

1. 11.64 m^2
3. 576 ft^2
5. 95 in^2
7. 0.16 in^2
9. 550 N/m^2
11. $\frac{9}{32}$ = .28125 in^2

13. a. $28\frac{15}{16}$ in^2
    b. 5.4"
15. 0.44 in^2
17. 1,078 ft^2
19. 7.31 lb/ft^2

21. 288 in^2
     806 cm^2
     11 yd^2
     20 yd^2
23. 19%

Unit 31: Volume and Mass Measurement

1. 56.6 in^3
3. 58.2 ft^3
5. 1 m^3 = 1×10^6 cm^3
7. 22.8 cm^3
9. 18,720 lb/s

11. 20,347.2 in^3
     11.775 ft^3
13. 37,324.8 g
15. 108 ft^2

17. 7×7×4 box
19. 132 refills
21. 47.7 L
23. 64

Unit 32: Energy, Work, and Temperature Measurement

1. 15° C
3. 77° F
5. 288 kelvins
7. 21° C
9. 35 N over 2.5 feet

11. 2,238 W
13. a. 18,000 kg-m/s^2 for A
    b. 36,000 kg-m/s^2 for B
15. 32° F; 273 kelvins
17. 3.77×10^6 Btu

19. a. 1,750 lb
    b. 583 ft-lb
21. 249 W

Unit 33: Measurements Involving Time

1. 42.3 mi/hr
3. 4.72 ft/s^2
5. 4 ft/s^2
7. 7.8 ft/s
9. 720,000 J or 720 kJ

11. a. 12:00 noon
    b. 11:15 a.m.
13. 12.44 minutes
15. 95.3 ft/s = v
    15.4 ft/s^2 = a

17. a. 11,200 J
    b. 0.6 hp
19. 2.875 ft/s
21. 478 rpm

Unit 34: Review and Combined Problems on Measurement

1. 8.75'
3. 90.322 cm^2
5. 88.495 km/h
7. 121.12 L
9. 1.832 radians
11. 2,154.504 N/m^2

13. $188.75 total
15. 5¼ in^2; 34.375% waste
17. 2,856 ft^2
19. 226 in^3
21. 1,928.88 g
23. 62 full days

25. 895,200 J
27. 63.75 ft-lb
29. 4,032 ft^3
31. 280 ft/m
33. 89.4 gal/min

Unit 35: Line Graphs

1.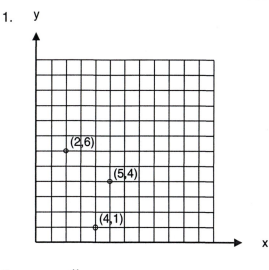

3. B = (4.5,1)
    C = (4.5,2.5)
    D = (3.5,2.5)
    E = (3.5, 3.25)
    F = (2.75, 3.25)
    G = (2.75, 2.75)
    H = (1.75, 2.75)
    I = (1.75, 2)
    J = (1, 2)
5. 5 hr
    5 hr
    6 hr
    6 hr
    4.5 hr
    5.5 hr

7.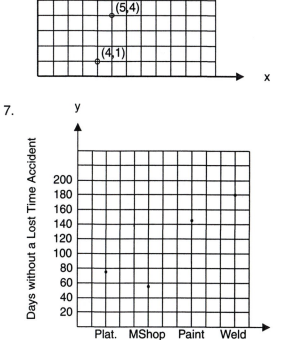

Unit 36: Pie Graphs

1. 0.27
   0.45
   0.08
   0.16
   0.04

3. a. 16 dot matrix
   b. 23% laser
      31% inkjet
      46% dot matrix

5. 79.2
   172.8
   72
   36
   total = 360

7.

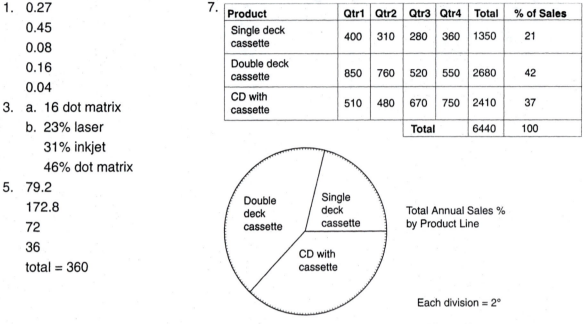

Product	Qtr1	Qtr2	Qtr3	Qtr4	Total	% of Sales
Single deck cassette	400	310	280	360	1350	21
Double deck cassette	850	760	520	550	2680	42
CD with cassette	510	480	670	750	2410	37
				Total	6440	100

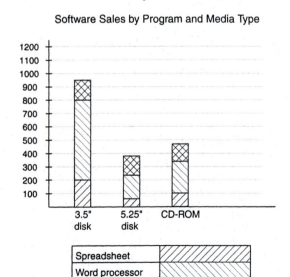

Total Annual Sales %
by Product Line

Each division = 2°

Unit 37: Bar and Stacked Bar Graphs

1. 700
   400
   600
   300

5. a. 3:1
   b. vertical scale
   c. 4:3

3.

Software Sales by Program and Media Type

Spreadsheet	
Word processor	
Database	

Software Sales by Program
and Media Type

Unit 38: Representation in Formulas and Equations

1. $P = E^2 \div R$
3. $e = E_o/E_i$
5. $r = (½)d$ or $r = d/2$
7. $c^2 = a^2 + b^2$
9. $R_T = R_1 + R_2 + R_3$
11. $e = 100(\Delta l \div l_i)$
     $\Delta l = (l_f - l_i)$
     $e = 100[(l_f - l_i) \div l_i]$

13. Amperage is equal to voltage divided by resistance.
15. Kinetic energy equals ½ of the product of the velocity and mass.
17. Gross salary is equal to the number of hours worked multiplied by the hourly rate of pay.

19. The temperature in degrees Celsius is equal to the product of ⅝ and the difference between the temperature measurement in degrees Fahrenheit and 32.

Unit 39: Solving Equations

1. $E / I = R$
3. $P / I = E$
5. $v_f = at + v_i$
7. $F = ⅗C + 32$
9. 20.2 ohms

11. $123.00
13. 21° C
15. 29.15 ohms
17. 13 teeth

19. 660.4° C
     1,984.8° F
     1,064.4° C
     621.3° F
21. 57.3°
23. 8 ohms

Unit 40: Formulas Common in Industrial Technology

1. 1,527.88 rpm
3. 2⅔ bd ft
     5 bd ft
5. 900 rpm

7. 121 ohms
9. 4.6
11. 149.2 W
13. 54 in^3

15. 7.3 ohms
17. 0.36

Unit 41: Pythagorean Theorem

1. 1.414
     3.09
     8.6; 26.0
3. 47 lb

5. $x = 1.4"$
     $y = 0.8"$
     $z = 1.0"$

7. 14.4 cm
9. 266 N

Unit 42: Trig Functions

1. a = adjacent
    b = opposite
    c = hypotenuse
3. 41.4°

5. 65.6°
7. 44.4°
9. 18.8'
11. 1.94"

13. 29.7°
     4.84'